War and Political Theory

And Political Theory series

Hawkesworth, *Gender and Political Theory*
Orend, *War and Political Theory*

War and Political Theory

Brian Orend

polity

First published in 2019 by Polity Press

Polity Press
65 Bridge Street
Cambridge CB2 1UR, UK

Polity Press
101 Station Landing
Suite 300
Medford, MA 02155, USA

ISBN-13: 978-1-5095-2496-9
ISBN-13: 978-1-5095-2497-6(pb)

A catalogue record for this book is available from the British Library.

Library of Congress Cataloging-in-Publication Data
Names: Orend, Brian, 1971- author.
Title: War and political theory / Brian Orend.
Description: Cambridge, UK ; Medford, MA : Polity Press, 2019. | Series: And political theory | Includes bibliographical references and index.
Identifiers: LCCN 2018039928 (print) | LCCN 2018056307 (ebook) | ISBN 9781509525003 (Epub) | ISBN 9781509524969 | ISBN 9781509524969(hardback) | ISBN 9781509524976(pbk.) | ISBN 9781509525003(ebook)
Subjects: LCSH: War (Philosophy) | Politics and war. | War--Moral and ethical aspects. | Just war doctrine.
Classification: LCC U22 (ebook) | LCC U22 .O638 2019 (print) | DDC 355.0201--dc23
LC record available at https://lccn.loc.gov/2018039928

Typeset in 11 on 13 pt Monotype Bembo by Servis Filmsetting Ltd, Stockport, Cheshire
Printed and bound in Great Britain by CPI Group (UK) Ltd, Croydon

For further information on Polity, visit our website: politybooks.com

Contents

Acknowledgements   vi

Introduction   1
1   The Ontology of War   6
2   Realism: Power, Security, and Nationalism   32
3   Pacifism: Ethics, Cosmopolitanism, and Non-violence   56
4   Just War Theory and International Law: Start-of-War   80
5   Just War Theory and International Law: Conduct-
    during-War   111
6   Just War Theory and International Law: End-of-War   138
7   The Future of Warfare   165

Notes   186

Acknowledgements

Thanks so much to everyone at Polity for helping to bring this book into being. Special thanks to Julia Davies, and extra-special thanks to George Owers, for the initial invitation to join the series and for very helpful, constructive editing efforts. Plus, for patience! Thanks to Neil de Cort for his production expertise, and also to Adam Renvoize for the excellent cover. Thanks so much, too, to my anonymous reviewers and to my copy-editor Ian Tuttle: the result of all the excellent feedback is a much-improved book.

A bit of an unusual dedication but sincerely expressed: I'd like very much to thank everyone who's ever invited me to come to their institution to talk about war and peace. Such invitations have been amazingly generous and the source of much stimulation and happiness for me over the years. Thanks again, and here's to the continuing conversation!

Brian Orend
August 2018

Introduction

Our world can seem in constant conflict. Such struggles range from bullying and bitter Twitter debates all the way up to crime, terrorism, and war. War is, historically, one of the most impactful forms of conflict, shaping all our lives. Even if you haven't been personally involved in war – whether through fighting or fleeing fighting – it has been involved with you. The social institutions of your country, for example, have been deeply determined by the wars of the past. This may have decided the very language you use to speak and think, and the population of people from which you've chosen your friends and loved ones. Many of today's most prized pieces of technology – notably, the Internet and mobile phones – began as military inventions. Into the future, the realities of the Global War on Terror (GWOT), campaigns of cyber-harms, and the rivalrous schemes of such countries as America and Russia, as well as China, Iran, Israel, and Saudi Arabia (and let's not forget all those armed non-state actors like al-Qaeda and ISIS) will configure some of the contours of our common life.

This book strives to widen and deepen reflection on three crucial questions about war: What is war? What should we do about it? And: How will war unfold in the future? In other words, this book aims to both *describe and analyze* war as a vitally important factual phenomenon, both now and into the future, as well as to evaluate war from a *prescriptive or normative* point of view. Ideally, this book aims to do so in a way both accessible and instructive for those with minimal education about armed conflict, yet will still stimulate worthwhile thinking, and rewarding questions, even amongst those with advanced background in the subject. The ambition is genuinely interdisciplinary, drawing heavily on facts and real-world cases, both historical and (mainly) contemporary, as well as such diverse theoretical resources as applied ethics, feminism, military strategy, philosophy, political science, and international law. The approach here is *less* one of arguing in favor of a particular theory or perspective and *much more* one of crafting an excellent, detached understanding of the pros and cons of the most important theories, and the most meaningful factual contexts which make the theories come alive.

Though the ontology of war is, inescapably, present throughout the whole book – "ontology" meaning the nature or reality of something – it will inform chapter 1 especially, and be rejoined at the end in chapter 7. It is pivotal to *define and examine* war in its manifold forms – e.g., civil versus international, symmetrical versus asymmetrical – alongside its various actors, and their means and ends of action, prior to a rewarding discussion of what we should do about it. There is not much point making judgments about, or recommendations for, something without knowing its form and nature, and how these evolve through time.

When we turn to such value judgments – in chapters 2–6 – we shall do so while drawing most heavily on what have been, and remain today, the three most foundational, detailed, influential, and richly suggestive theories. They are on a logical continuum, as shown in table I.1.

One such theory states that *we should never go to war*, especially when we know from history how devastating it is, how costly, brutal, and painful it is, and how frequently wars do not unfold the way one hopes or predicts. And there are compelling moral reasons, and abiding ethical principles, as to why we should not deliberately kill our fellow human beings, much less on the mass

Table I.1 Continuum of Foundational Theories about Warfare

Realism	JWT/LOAC	Pacifism
Pessimistic extreme	Middle ground	Optimistic extreme
About war: *"anything goes"*	About war: *"sometimes, something goes"*	About war: *"nothing goes"*
One large internal division: Classical vs. Structural realism	One large internal division: Traditional vs. Revisionist JWT	One large internal division: Religious vs. Secular pacifism

JWT, just war theory; LOAC, laws of armed conflict

scale (and for "merely political" reasons) demanded by warfare. This, of course, would be the doctrine of **pacifism**, and it occupies what is arguably the optimistic and idealistic extreme in this regard: there's always a superior option to war-fighting, and in terms of violent armed conflict, *nothing is permissible* and war should be outlawed and banned. A simple slogan of pacifism, in connection with war, might be: "Nothing goes!"

At the other extreme, the bleak and pessimistic one, would be **realism**. Realists tend to view all of human relations – or, at least, all of international relations – as governed by an endless power struggle. War is an entirely predictable consequence of this fact, as it's one major method humans have to try to gain and hold power. There might not even be much we can do about it: there's certainly no historical reason for thinking we can successfully ban warfare, or do away with it completely, as we all march together peacefully into a sun-lit future. To the extent to which we *can* do something about armed conflict, it should be to try to ensure that, when we get involved in warfare, we do so *only because* it's manifestly in our self-interest (i.e., that it's worth our while in terms of cost–benefit analysis) and, once we're in it, we should be in it to win. So, anything conducive to the end of victory is fair game, and should be considered according to the same self-interested cost–benefit calculation as the original decision to go to war. Thus, the simple slogan capturing the essence of realism, in connection with war, might be: "Anything goes!"

To complete our basic logical continuum: if one option is to let anything happen in connection with war, and just leave it to the vicissitudes of perpetual power struggle; and if another option is to condemn all warfare absolutely, and assert that it's never the proper thing to do, then of course that would leave a middle-ground option, according to which *sometimes* war might be both wise and even morally permissible, whereas other times it's ill-considered and even unethical. This is the core proposition of just war theory or **JWT** (sometimes, war might be morally justified). It's also an axiom of the so-called "laws of armed conflict" (**LOAC**) which have been agreed upon by most state governments around the world as one form of international law. Though the relationship between JWT and LOAC is not totally straightforward, as we'll see and explore, it *does* make substantial sense to suggest they are close conceptual allies in this regard.

JWT/LOAC is thus the most complicated of the three basic doctrines or theories, as it takes quite a bit of time, effort, and complexity to specify which exactly are the times when war might be smart and just, and which other times it's unwise or unjust. Realism and pacifism are more streamlined and straightforward, as they are more conceptually "pure" and one-sided in their attitude and understanding. This has the virtue of making them easier to understand and develop, and to see their forceful strengths, but may come at the cost of being more extreme and, in the end, harder to endorse overall. We will see that JWT/LOAC has substantial problems of its own. Alas, *the perfect perspective on warfare may yet to have been invented.* But such realities explain the breakdown of this book: after some good definitions and salient facts, we can orient ourselves quickly and powerfully by looking at the basic theories at either extreme of opinion. Realism thus gets chapter 2, and pacifism chapter 3. Then, we need more time and effort to explore the complex middle ground which, in a sense, attempts to split the difference. So, chapters 4–6 will analyze and critically evaluate JWT/LOAC, applying it fruitfully to many concrete cases.

We won't ignore other theories – such as anthropology, Freudian psychology, "IR theory," and the democratic peace thesis – and will draw upon them when the time is right. Moreover, we'll pay considerable attention to *the internal pluralism, and issues of division, within* each of The Big Three theories. There are a number of such internal divides, and some of the most elemental are charted in Table I.1. In any event: the heart and soul of this book, when it comes to the issue of what to do about armed conflict and warfare, is to explore in rigorous, colorful, factually informed fashion, the powerfully stimulating and instructive debate between realism, pacifism, and JWT/LOAC, alongside the internal debate each tradition has regarding how best to view and apply its own core ideas and values. Out of such a spirited conceptual clash comes deeper theoretical understanding as well as a greater practical grasp of what we might do about the specter of constant conflict.

1

The Ontology of War

Ontology, as we've seen, means the being, nature, or reality of a thing. Our investigations into war's ontology in this chapter will thus involve definitions, components, key players, and theories regarding the ultimate causes and origins of warfare.

War is well defined, at least initially, as an actual, intentional, and widespread armed conflict between groups of people. This is true whether these groups are *within* one country (civil war) or in *different* countries (international warfare).[1]

Nearly every country in the world has suffered from a civil war at some point in its history, often with profound impact on its future. There is a civil war raging today in Syria, for example, between the government and those who wish to overthrow it. This brutal conflict has destroyed entire cities, killed nearly half a million people, and sent millions more fleeing for their lives as refugees into countries as diverse and far-flung as Turkey, Germany, Canada, Lebanon, and Australia.[2]

There are of course thousands of examples of international warfare between countries, perhaps most impactfully the two world wars: World War I (1914–18), which saw about 20 million people killed, and the largest single war in human history, World War II

(1939–45), which involved more than 50 countries and saw at least 50 million people killed (recent research suggests the number may actually be more than 70 million). Both conflicts had profound effects on the world. New weapons were invented, ranging from the tank to nuclear bombs; the modern Middle East was deeply shaped, both for good and ill; the old empires of Europe (especially the world-wide ones of Britain and France) began to fall apart forever; discriminatory, undemocratic, mega-violent regimes in Germany, Italy, and Japan were destroyed; and the two largest winners of the second war, America (USA) and Soviet Russia (USSR), immediately plunged into a world-wide Cold War between themselves, which was to last for the next 50 years, taking us into our own time.[3]

1.1 The Elements of War

Let's analyze further the elements within our basic definition of war. A war, whether civil or international, must be *actual*. This is to say that there must be actual fighting, and the deployment of armed force and violent physical attacks – as opposed to mere threats, angry talk, a military build-up along the border, and even policies like economic sanctions, which are blockades or bans on commercial trade with a target country. All these other measures, while no doubt intended to influence (or manipulate, or even coerce) the target country or group, do *not* count as acts of war. *Logically, it simply must involve violence* – physical force, designed to harm and coerce the other[4] – *for it to count as war.* Sometimes today the term "hot war" is employed in this regard: an actual, violent, shooting war between the opponents, involving killing, injury, property destruction, and such like, as opposed to other terms, like "phony war" or most clearly "The Cold War." The "phony war," for example, was a period very early on in World War II, after Nazi Germany had invaded Poland in 1939 – triggering declarations of war from Britain and France – but before any further fighting happened, which came several months later in 1940, when the Nazis launched their blitzkrieg ("lightning war") to very rapidly invade and conquer France and Holland, serving up a stunning initial setback for the Allies.[5]

The Cold War, as mentioned, refers to the complex, multifaceted world-wide struggle – from the end of World War II in 1945 until approximately 1989–91 – between the two largest winning powers of that war, namely the US and the USSR. These two "super-powers" each represented totally different forms of political and socio-economic organization: with America being the leader and champion of democratic, free-market, liberal capitalism and Russia serving as the symbol of undemocratic, centrally planned, "scientific" communism. As these countries were both so powerful, and in particular had developed such enormous arsenals of nuclear weapons, they really couldn't dare to go to war against each other, lest mutual destruction occur. But they were bitter enemies, committed to the defeat and collapse of each other's political values and social systems. So, it was a "cold war": no direct violence (or hot, killing war) between the two, yet still a committed and costly multi-generational effort, using a variety of methods, to bring about the defeat of the other. Eventually America won, and the Soviet Union collapsed, with such symbols of the Cold War as the Berlin Wall crumbling in 1989, and elected governments coming into power throughout Central and Eastern Europe in the early- to mid-1990s. It is sometimes jokingly said that the Cold War is the only major war with zero casualties, yet global consequences.[6]

But that's not quite true, actually: the "zero casualties" part. For, even though the USA and USSR could never afford to fight each other directly in a hot war – lest it escalate to the point of world-wide nuclear devastation – they each nevertheless funded and armed, and manipulated and allied themselves with, various groups and countries around the world, who *themselves* engaged in hot wars. This phenomenon is important, and it's called a "proxy war." This is when powerful rival countries support other, lesser players, and essentially get them to "fight their battles for them," as a way to harass, impose costs on, and thwart the agenda of the other powerful rival. The Cold War was filled with many proxy wars, which generated millions of deaths and injuries, and affected the fate of regions as diverse as Southeast Asia (e.g., Vietnam), the Caribbean and Central America (e.g., Cuba and Nicaragua), and sub-Saharan Africa (e.g., Mozambique and the Congo). Proxy wars endure today; one of the reasons why Syria's civil war has been raging for so long is because regional powers Iran and Saudi Arabia – bitter rivals within the Middle East, for ethnic, cultural, and religious reasons – have treated it like a proxy war between themselves, each arming and enabling different local actors or belligerents (i.e., those who do the fighting in a war).[7]

Back to the elements of our initial definition of war: the armed conflict in question must not merely be actual, but also *intentional* and *widespread*. The belligerents must, in some robust sense, truly *want* to go to war with each other, and to cause each other substantial harm. Traditionally, this has been signaled through things like a formal, public declaration of war by some official government body, be it the head of the executive branch (usually, the head of state such as a president) or, instead, a majority vote in the legislative branch (like a parliament or congress). While such proclamations *do* signal hostile intentions, they *don't* of themselves count as war: witness "the phony war" above, for example, wherein there had been declarations but no actual fighting yet. The reverse can be true, too: deliberate, war-scale violence has been known to occur in the absence of such public declarations (like the early stages of the Vietnam War, until the Tonkin Gulf Resolution in 1964, or many proxy wars, such as in Syria, wherein Iran and Saudi Arabia have never declared war on each other). The actual behavior is the crucial thing, or test; and thus, many experts think that we should see the "deliberate intentions" part going together with the "widespread" part as an indicator of actual war: presumably, there cannot be widespread exchange of "hot war" physical violence without the belligerents sincerely wanting that to happen, and setting all that into motion. This may be important to distinguish a genuine war from things which might *look* like war (or are indicators of *imminent* war) yet which aren't.

Consider that sometimes some fighting can break out between isolated military units who are on patrol along a tense international border. There were reports of fisticuffs between border patrols along the India–China border in the summer of 2017, for example, and sporadic shots are quite regularly fired across the India–Pakistan border. We can see how such can readily happen: a few hot-headed words occur, someone gets angry and fires a shot across the border, return shots are fired, there might even be a casualty or two before the heated exchange ends. But we wouldn't say that, therefore, *there's now a war* between these two countries. Sometimes the chain of command breaks down, and there are some soldiers, or even whole units of soldiers, which in the heat of the moment have been known to do their own thing, even in defiance or violation of their own government's orders. So, here's a clear illustration of how the requirements of deliberate intent and widespread violence *go together*: such border skirmishes do not show the intent of the respective national governments to go to full-blown war against each other, and the killing isn't widespread enough for us to clearly diagnose a war in action. There needs to be *a certain quantum of force*, and killing, for us to talk about war.

Now: *what* quantum, exactly? How many people need to be killed before we can talk of war? Social scientists for a long time have used the figure and language of "1,000 battlefield casualties" (i.e., deaths and severe injuries) to diagnose when a war has actually broken out. On the one hand, this number seems arbitrary: why not 2,000? Why not 800? On the other, the number was selected as a serious standard to distinguish between a true *war* between determined belligerents who are "in it" for the long haul and, say, a large-scale police action (like a raid on a criminal gang's headquarters, which produces 25 deaths in a shoot-out); or an urban riot between different private groups of people (such as in times of extreme racial or ethnic tension), resulting in the deaths of 47 people over the course of a hot summer weekend. One thousand is also a number which can actually be counted, and therefore is empirical and evidence-based. The standard has also been chosen because 1,000 seems to reveal that element of hostile intention: presumably, so many people cannot be killed or severely injured by violence unless the belligerents are truly committing to the significant and ongoing use of physical force to try to achieve whatever objective they have in mind. And why "battlefield" casualties? Precisely to distinguish between a war and other mass-casualty events like earthquakes, famines, epidemics, some terrible accident like a plane crash, and so on. And don't let the "battlefield" induce too many historical images. The term is still used, not so much to refer to so-called "set-piece battles" on an old-school battlefield – such as the Battle of Waterloo in 1815, which saw the final defeat of Napoleon – the language is retained to show that *it must be deliberate fighting between belligerents who genuinely want to go to war against each other,* whether such is in a field, in the jungle, in underground cave networks, in dense urban environments, or indeed on the water or in the air.[8]

There's further value in something like "the 1,000 battlefield casualties" standard: if we apply it, we see that there have been more than 200 wars (including both civil and international wars) in the last 100 years alone. So, *on average, there are two new wars every single year:* a fact which is important to know and may have serious implications for how we think about politics, both domestic and international, as well as the use of deliberate, killing violence as a tool in pursuit of a political agenda. At the time of writing in 2018, there seem to be, around the world, about 14 genuine wars ongoing – and perhaps close to 50 more if one drops a bit below the 1,000 casualties standard. War, clearly, is a grave, regular, and ongoing feature of human history, shaping nearly everyone's experience and life in some way, either directly or indirectly. This naturally raises questions as to whether anything can, or should, be done about it, which forms an undercurrent to this entire book.[9]

The next element of our basic definition of war as *an actual, intentional and widespread armed conflict between groups of people* is the "armed conflict" part. By this, we mean the use of weapons and physical violence with the intention of inflicting damage and harm upon people in an effort to get them to do what you want. About weapons: more in a minute. Ditto for "groups of people." For now, consider the thoughts of one of the few so-called "philosophers of war," Carl von Clausewitz. Clausewitz was both a soldier and a strategist. He was a Prussian (East German) officer during the Napoleonic Wars (1800–15), and over 16 long years wrote a landmark book called *On War,* still studied at today's military academies around the world. It's mainly about military strategy but also contains meaningful philosophical reflections on the essence of war. His most famous quote is that "war is the continuation of politics by other means," commonly interpreted to imply that war is just another tool people use to try to get what they want, and that warfare centrally revolves around the distribution of power between groups of people. We'll come back to such notions. For now, note Clausewitz's own definition of war: "an act of violence, intended to compel our opponent to fulfil our will." War "is like a duel," he concludes, "only on an extensive scale."[10] Such an account seems consistent with what we have offered here.

Before analyzing the "groups" component of our initial definition of war, there are two further aspects to mention, before leaving them on the table for deeper consideration at the right time in later chapters. In chapter 4, we'll delve into the LOAC's definition of war, which revolves in the first instance around the use of physical force across a border in violation of state sovereignty. The thick concept of "aggression" is crucial here, and raises complexities regarding civil wars and other phenomena – like genocide and ethnic cleansing – which look an awful lot like war yet which may not spill over an international border.

In chapter 7, we'll consider how the rise of advanced computer technologies has enabled malign, harmful cyber-attacks between groups. Is cyberwar truly war? Obviously, cyber-tools can *complement* and *further* conventional, physical war as we've defined it. But does cyber also mark a unique kind of war, consistent with our initial definition? There's a spirited debate about this which we'll rejoin at book's end, as we ponder whether the nature of warfare itself is changing as we move into the future.

1.1.1 Which Groups?

We've already distinguished between two kinds of groups which might get involved in warfare: competing groups within one country, sparking civil war, or else two or more different countries, in classic international warfare. An important distinction used today is between so-called symmetrical and asymmetrical warfare.

Symmetrical warfare refers to "old school," government versus government, warfare, like the world wars: epic clashes between official national militaries. ("Symmetry" in the abstract sense of "balanced," "one-to-one," or – most accurately – "between the same sorts of thing.") More recent cases of symmetrical warfare would include the two wars involving Iraq. First there was the Persian Gulf War of 1991, to drive the Iraqi military out of neighboring Kuwait, which it had invaded and conquered the previous year (mainly to get Kuwait's oil, but also because Iraq's President, Saddam Hussein, wanted to give his huge military something to do, following its displeasure at his having called off Iraq's war with despised rival Iran the year before). The United States led that quick war – successful in the sense of ejecting Iraq from Kuwait – and was joined by nearly 30 allied countries from around the world (including Arab countries like Jordan and Saudi Arabia). This was quite unlike the case in the second war, in 2003, when the US only received the support of the UK, as for various controversial reasons George W. Bush ordered the invasion of Iraq, and the forcible removal of Saddam's regime, resulting in prolonged and problematic US occupation of that country.[11]

Asymmetrical warfare refers, by contrast, to armed conflict between a government and various "non-state actors" (thus, "asymmetry" in the sense of "between different kinds, or levels, of thing"). The official government of a country, we know, is that country's "state." Thus, a non-state actor refers to an organization which is not in any way part of the official government of that country. Non-state actors are of increasing importance in today's politics, both nationally and internationally; the most influential of them range from business corporations to media sources (both old and new), and from religious associations (like churches) to non-governmental organizations (NGOs) focused on performing a specific task, often on a not-for-profit basis and focused on important social issues like protecting the environment (Greenpeace), human rights (Amnesty International), or emergency medical care (Doctors Without Borders).[12]

Alas, not all non-state actors are peaceful, nor are they all selflessly focused on humanitarian causes which benefit the world.

These are the sorts of non-state actors typically involved in asymmetrical warfare. They could include:

- *Revolutionaries or insurgents*: i.e., those trying to overthrow the official government of a country, and replace it. Examples range from those sparking the French Revolution (1789), the Russian Revolution (1917), and the Chinese Revolution (1949), to Iran's more recent revolution (1979), wherein the old "Shah" or king was overthrown and replaced by a regime committed to the use of state power to realize a strict religious vision, in this case that of an immoderate understanding of Islam. This regime, founded by Ayatollah Khomeini, is still in power today.[13]

- *Armed separatists* who wish to leave a particular country, taking some territory and population with them, and forming their own, new, independent country. A classic example would be the American revolutionaries who wished to secede from the British Empire and form their own new country, sparking the US Revolutionary War from 1775 to 1783. Perhaps the largest group of people right now like this are the Kurds, an ethnic group whose traditional territory overlaps with the existing countries of Turkey, Iraq, Syria, and Iran. The Kurds are often referred to as the world's largest "stateless nation," and they have taken up arms on a range of fronts – against each of the governments above – in search of the establishment of their own new, separate country, Kurdistan.[14]

- *Terrorists*, such as al-Qaeda or ISIS. Terrorism, strictly, is a tactic, and not a full-blown "ism" (i.e., a political ideology, or detailed conception of social justice like liberal capitalism). Terrorism refers to the use of random, deadly violence – especially aimed at the non-military, civilian population – with the hopes of spreading fear (terror) throughout that population, believing that such widespread fear will advance an agenda which the terrorist group has. So, al-Qaeda attacked America on 9/11 in 2001, killing thousands of civilians, with the goal of having the American people pressure the US federal government to get out of Middle Eastern politics, thereby strengthening the hand of radical Islamic groups like al-Qaeda (so that they might be able, for example, to overthrow the government of Saudi Arabia – long armed and supported by the US – and establish an Islamic state there, in the holy land of Islam). Or consider all the recent truck attacks on civilian tourist centers in Western Europe, by ISIS, wherein the intended goals of such strikes might include: (1) harming a society they see as utterly opposed to their values; (2) getting revenge on a country whose military is involved in harming their interests in the Middle East; and/or (3) using such attacks for propaganda purposes to recruit new members and to inspire existing members.[15]

- *Private military companies (PMCs)*: These are better known, historically, as mercenaries: i.e., people who fight in war merely for personal wages and not for any moral cause, political reason, or national loyalty. Mercenaries actually have a very long history within warfare (e.g., many of the battles in the American Revolution were actually fought between the American revolutionaries and German mercenaries from the principality of Hesse, hired by the King of England). Today's PMCs hate being called mercenaries, yet the fact remains that they do what they do for the sake of money. Most PMC staff are ex-military, who've moved over into PMCs in search of higher wages: they tend to be viewed dimly by those still inside national military establishments. There are many PMCs today in sub-Saharan Africa, and they are often employed by national governments as a way to minimize the involvement of their official militaries (thereby keeping down unpopular things, such as "the body count" of their own official soldiers). During the recent postwar occupation of Iraq, the US government hired a number of PMCs – such as Blackwater (now called "Academi") – not so much to fight on its behalf but, rather, to do things like hold and help police an Iraqi town after the US military had won it in battle against insurgents and needed to move on to the next. We might see how that *could* be an important and understandable service, deserving of reward, and yet questions remain about accountability and legitimacy when it comes to authorizing the use of potentially deadly force by a non-state actor, much less one devoted to private profit.[16]

- *Well-armed and violent criminal enterprises*, such as drug cartels: Recent examples would include Colombia and the Philippines, but especially Mexico, wherein the national government is pitted in a violent, ongoing conflict with gangs which smuggle people and drugs into the United States, and the scale of such violence over recent years has reached near-warlike intensity and casualties. In some regions of Mexico, it's actually not clear whether the government actually has control and the ability to govern, as compared with the well-resourced, well-armed, and utterly vicious drug lords. The situation certainly blurs the line between a police action and a military engagement.[17]
- *Self-styled "militias,"* a term used as a catchall phrase for any non-governmental armed group. Most commonly, these militias tend to see themselves as having a political cause or purpose (like preventing the national government "from becoming a dictatorship") and thus are neither mercenaries nor drug gangs. They might "talk-the-talk" of something which sounds like revolution, or secession, but generally they just seem like angry men with guns, keen on supporting some particular policy, such as their right to own guns. Fans of militias will point to their origins in things like the American Revolution, whereas skeptics will observe that, the more angry men with guns there are, the more likely it is that acts of serious violence will follow.[18]

Note how some of these categories may and do overlap: drug lords may employ mercenaries, for example, or militiamen become revolutionaries, or separatists may turn to terrorism as a tactic in their ongoing struggle. Note – moreover and importantly – how *so many of today's wars tend to be complex blends of both symmetrical and asymmetrical elements*: involving governments both local and foreign; official national militaries; aspects of civil war between local groups; militias, insurgents, secessionists, and mercenaries; criminals trying to capitalize on the chaos of war; radical ter-rorist groups; and various agencies (whether state or non-state, whether nearby or far-away) providing assistance to the various local belligerents. All vying to skew things in their own preferred direction. This complex fact is true, for example, of today's largest conflicts (in terms of casualties): Syria, Afghanistan, Iraq, Somalia, and South Sudan.[19]

1.2 Military Capability in General, Weaponry in Particular

War implies *armed* conflict between some mixture of such various groups. Let's therefore quickly define some helpful, much-used, military terms and concepts, and consider some initial things about weaponry, to which we'll return in chapter 5. We'll begin with the traditional set-up of a country's official military and move from there.

Military capability refers to the resources a country has to deploy armed force. Almost all countries have an army, a navy, and an air force, each favoring its own kind of weaponry, ranging from tanks to ships to fighter jets, respectively.[20]

More recent additions to this core trinity – army, navy, air force – of military capability are cyber-operations and special forces. Cyber-operations are about trying to control and defend "cyber-space" – not just the Internet but, really, anything related to the functioning of crucial computer systems – in accord with that country's national interests. Special forces exist to execute unusual, very specific military objectives, which usually fall under the rubric of "measures-short-of-war" (MSOW) but still involving the projection of professional, potentially lethal, armed force. Special operatives in the US military, for example, include the Green Berets, Navy SEALs, and several branches of the Marines. Much of their work is highly classified. Counter-terrorism has been a large part of recent missions: a high-profile instance of special forces activity was the manhunt for, and execution of, Osama Bin Laden in Pakistan in 2011. Such so-called "targeted killing" – perhaps otherwise known as assassination – is not usually considered an act of war, despite its violence, precisely because it is individually targeted and thus fails to meet the standard of being widespread.[21]

When working together during wartime, a nation's military

forces aim to control the air, water, and land of a given region, and to use the tools of special forces and cyber-operations to complement these core objectives. This would give that nation maximum *leverage* to inflict its will upon that region's population, enabling the fulfillment of its war aims.

Two further aspects of military capability must be stressed. The first is logistics, which concerns the ability of a country to supply its armed forces with everything they need to fight, which ranges well beyond weapons to include such things as food, clothing, medicine, and gasoline. Supporting a military is a massive effort of resources, planning, and organization. Historically, it has often been as damaging to strike an enemy's supply chain as to strike directly at its armed forces. Both Alexander the Great (356–323 BCE) and Napoleon (1769–1821) had to give up their campaigns of conquest because of severe logistical problems. The second aspect is intelligence, which entails the gathering and analysis of data about the enemy: its weapons and capabilities, its plans and intentions, its most vulnerable or most strategic targets, and so on. Most countries have some kind of intelligence agency to serve this function. In the US, for example, it's the Central Intelligence Agency (CIA); in the UK, the Secret Intelligence Service (SIS, still often referred to as "MI6"); and in Canada, the Canadian Security Intelligence Service (CSIS).[22]

Modern armed forces are thus amazingly expensive, both to assemble and to maintain. All the soldiers and sailors, and all the pilots and marines, must be fed, clothed, housed, trained, and paid an income; all the weapons must be built and kept battle-ready, and then replaced after they break down or become obsolete (i.e., out of date and inferior). Transporting troops and their weapons is difficult and costly, too, and the technology for doing so can range from simple Jeeps and Humvees to naval aircraft carriers, nuclear-powered submarines, cargo planes, and stealth bombers (which can avoid radar detection). And when it comes to weaponry, no one wants to be caught using yesterday's technology, which means that countries are continually paying to upgrade their equipment. (It's not just an issue of trends: military history often shows that weapons superiority determines victory or defeat.)[23]

1.2.1 Strategy versus Tactics

Like weapons technology itself, the study of how best to fight different kinds of enemy is constantly evolving. Military planners commonly distinguish between strategy and tactics. Strategy refers to the big-picture plan of how to defeat an enemy in war, whereas tactics refers specifically to how one should fight particular battles *as pieces in* the overall strategy. A rough example: during World War II, the Allied *strategy* for beating Hitler was to surround him with overwhelming force – Soviets/Russians from the east; the British and their empire from the north and west; the Americans from the south and west – in order to draw an ever-tightening circle around him, until he either surrendered or was destroyed. This strategy of encirclement is as old as warfare itself, and was decided upon by the Allies in very short order. The *tactics* for carrying out this strategy varied from battle to battle, and took years to execute, notably including the successful "D-day" invasion of Normandy in 1944, and then the subsequent moves to roll the Nazis back within the borders of Germany and, by early 1945, for the Americans and Russians to "Race to Berlin," squeezing it from both west and east until the Nazis collapsed, Hitler committed suicide, and Germany surrendered unconditionally.[24]

Wealthy nations have a clear advantage when it comes to armed force, owing to the immense costs of military planning, weapons research and purchasing, intelligence gathering, and near-constant technological change in the field. But two things to note: first, too much military spending is bad for a country's economic growth over the long term (as the Soviets discovered by the end of the Cold War, and as is true for North Korea today); and second, enemies using unpredictable, low-level, counter-conventional "guerrilla tactics" can be capable of inflicting surprisingly large damage even on the richest and most powerful countries (witness the 9/11 attacks on America, or the recent terrorist truck and knife attacks in Western Europe). Recall that experts distinguish between symmetrical warfare, which is classical, open, direct warfare between states or between evenly matched sides, and asymmetrical warfare, which is warfare pitting a state government on one side against a non-state actor (like a terrorist or insurgent group, or several) on the other. As a rule, non-state actors do *not* fight the same way that states do, as they almost never have a comparable level of military capability. States tend to fight conventionally, as outlined above; non-state actors, by contrast, deploy unconventional warfare. They prefer sneak attack, deception, hidden weapons, "strike-and-melt-away," blurring the line between themselves and civilian populations: anything to reduce the large advantage which state governments have when it comes to general military capability. In particular, non-state actors almost never want to "come out into the open" and have a set-piece battle with any given national military: they would almost certainly be promptly surrounded and destroyed. They prefer to fight like "guerillas," a term coined by the low-tech yet inventive and utterly determined Spanish insurgents against, and resisters to, Napoleon's massive, well-oiled "Grande Armée," during France's invasion and occupation of Spain in the early 1800s. As a general truism, *the strategy and tactics deployed will vary substantially, depending on which kind of war, and which kind of adversary, one is fighting.*[25]

1.3 What Causes War, and Is There a Solution?

We have defined war as *a violent armed conflict between groups of people which is actual, intentional, and widespread* – and have now offered fuller analyses of each of the component elements. Yet it naturally occurs to ask: why do people do this? What drives all these different groups to go to war: to exhibit such extreme, risky, bloody, and costly behavior in pursuit of their goals? Why have there been so many wars in human history, creating such enormous misery? Could we ever realize a human future without armed conflict? This is perhaps the ultimate issue for war's ontology: what causes war?

There are various influential theories, each backed by some evidence, about the origins of war. These are explained here at least provisionally: in many ways, this whole book will be a meditation on what causes war and whether anything can be done about it. Note that no one theory seems completely satisfying, and several (or all) may figure in some way into the explanation for the eruption of any given war. We ought therefore to avoid reductive, single-cause, conceptions in this regard, and realize that the multi-cause model (while no doubt more comprehensive) complicates attempts to solve, or even manage, humanity's frequent, and tragic, predilection for violent armed conflict.

1.3.1 Realism and the Quest for Power

All of the next chapter will be devoted in detail to the realist world-view. It is one of the most important and influential regarding armed conflict. According to the realist – exemplified above by Clausewitz – war is all about power (i.e., the ability to get what one wants), with military force being one of the most potent means for getting others to comply with one's will. War is about the use of violence to exert control, whether domestically over one's own people or internationally over foreigners. A variant of this account is "the hegemony/backlash theory" of armed conflict, according to which wars are produced by one country's pursuit of hegemony (i.e., strong influence over others) and/or by those resisting such aggrandizement. This is a popular explanation for many recent European wars – especially the world wars, sparked by Germany's quest for hegemony within Europe – and for many of America's recent wars, from the Cold War to Afghanistan and Iraq.[26]

As to why humans, or groups, are so focused on power, we'll leave the full account for the next chapter. Generally, it tends to be a conviction either that such is part of human nature (traditional or classical realism); or, rather, that the concern over power and security is forced upon us by the insecure, sub-optimal set-up of the international arena ("neo-" or structural realism). In particular, there is no effective world government, and so – when push comes to shove – countries are on their own when it comes to providing for, and protecting, their own people. So, war is produced either by a country wanting more power, or else because it's the last resort option which countries have when it comes to some issue which they deem vital to their national interests, whether it's fending off an attacker, trying to access vital resources, or trying to disarm and neutralize a determined enemy.[27]

It's interesting to note that, even though the realist view seems quite bleak, there's actually a hint regarding a possible solution to the problem of war, at least for the structural kind. For, if a truly effective world government could be established – one that could settle disputes between nations the way effective police and court systems settle disputes between individuals in well-run national societies – then we could have something like a reliable solution to the problem of war. The sizable challenge for this idea, however, is creating such a cosmopolitan institution in the first place.[28]

1.3.2 Idealism and the Clash of Values

Political scientist Samuel Huntington, in his prominent, contro-versial book *Clash of Civilizations*, is skeptical of claims that war is about power over others, and especially the common claim (discussed below) that war is about gaining natural resources. He argues, instead, that war is generated and fueled by a profound disagreement over ideals – in particular, competing cultural values about justice and how best to run a society. Consider, he'd say, World War II and the Cold War as prime examples, as well as the great many wars of religion – in nearly every region – throughout history. Using these ideas, Huntington predicted – following the end of the Cold War circa 1990 – that the next major armed conflict would be between Western civilization and Islamic civilization.[29] On the one hand, he was kind of correct; on the other, he said some contentious things in developing this thesis, particularly about Islam's propensity for violence. Perhaps a more neutral expression of this thesis occurs in David Welch's *Justice and the Genesis of War*, wherein Welch argues that, more than anything, it's clashing moral and political ideals which explains why people are willing to engage in something as dark and dangerous as war: people get "hot and bothered" about such values, hold them deeply, experience profound anger over their violation, and thus can even be willing to kill or be killed over such things.[30]

Suffice it to say, the only way to imagine a solution to war on these grounds would be if there came to be substantial world-wide agreement on basic moral and political values. Historically, of course, this is *not* to be expected – though some truly universal norms *might* be emerging, such as some values associated with everyone's individual well-being, like human rights and global public health.[31]

1.3.3 Nationalism and Elite Manipulation

This theory hinges on the notion that wars are waged between groups of people, and usually – in the modern world – these groups are nations. A nation is a group of people *which sees itself as a people*, and as having a deep identity unique from other peoples (often revolving around a complex bundle of traits like ethnicity, language, religion, shared history, similar values, and shared aspects of culture (or "way of life"), like diet, dress, leisure activities, music, literature, and art). A nation thus defined is separate from a state, which (as we saw above) is more narrowly and technically the structure of government over a group of people. So, there can be nations without states, like the Kurds, and there can be states which govern over many national groupings, such as the US, Australia, Canada, and the UK – any of the large "immigrant-receiving societies," historically. Nationalism refers to the tendency for every nation to want its own state, and has been a major cause for many modern wars. Witness, for example, the many "wars of independence" fought in the twentieth century between the major European imperial powers, from England and France to Portugal and the Netherlands, and many of their (now) former colonies around the world, ranging from Algeria and Angola to Vietnam and Indonesia. And then, sometimes, there were cascading wars *within* those societies, as groups-within-groups fought over control of the new country, or else seceded and created yet another. The armed proliferation of nations, so to speak, is a compelling way to view much of international political history over the past 200 years.[32]

Nationalism, and a sense of national belonging, are very meaningful to many people in modern history. Given the power of the nation-state over so many aspects of social life, we've all been deeply shaped by the ideals and interests of our country. Thus, it's very plausible to suppose that such a strong identification with one's group plays a huge role in the outbreak of warfare, and in the ability of a government to sustain a war over time.

That adds a further dimension. This theory actually has two parts:

1. that national identification – i.e., patriotism – is a common, powerful reason why *ordinary people* agree to fund and fight wars; and
2. that *ruling elites* use, or manipulate, this prevailing sense of patriotism for their own self-interest in holding onto, or extending, their power.

This third theory is thus a blend (of a kind) between realism and idealism, as just described: the people fight for their values and ideals, whereas the rulers orchestrate the fight for their own benefit. Many a ruler has, historically, found war with a foreign power a quite convenient way of distracting his people from problems at home. Consider, for example, Saddam Hussein's 1990 decision to invade Kuwait. He drew heavily upon Iraqi nationalism, claiming that Kuwait was properly part of Iraq. The reality was that he had just finished a terrible 10-year war with Iran, which left Iraq enormously in debt. Moreover, he had a huge, restless, embittered military on his hands – a military which, historically, had determined who rules Iraq. The perfect solution for him was to invade Kuwait. This would keep his military busy and off his back, and it would give him Kuwait's oil wealth, which he could use to pay down Iraq's debt. (He did not expect that the US would care enough to get involved, thus beginning a sad series of severe miscalculations.[33])

It's hard to see a solution to war, if this blended account is its ultimate cause: group membership and patriotism can be intense, encompassing emotions, and not many rulers have been eager, historically, to lose power, or to rule for the main benefit of others. I suppose one issue is why national tribalism would need to find expression in mass violence – a consideration to which we'll return below. Suffice for now to say that it may, at least, serve as some kind of check or obstacle to armed conflict for ordinary people to view with healthy skepticism the designs of anyone with ambitions to lead them, and emphatically so when they wish to lead them into the bloody maw of war.

1.3.4 Money and Materialism

Vladimir Lenin, leader of the world's first communist revolution – in Russia, in 1917 – argued in favor of this theory. As soon as his communists prevailed in the revolution, creating the Russian Soviet Socialist Republic, he pulled that country out of World War I (WWI), leaving the others to fight it out – and he soon had his own problems (a civil war) to deal with. For Lenin, countries fight over the means of economic production, specifically land and all the natural resources it contains, both on top and underneath: people, animals, timber, crops, water, oil, gold, silver, coal, diamonds – you name it. Lenin argued that, as long as you have an economic system based on free-market capitalism, wars will break out, sparked by greed and competition over material resources. He blamed pretty much the whole history of European colonialism and imperialism, and certainly the outbreak of WWI, on this fact.

Lenin thought the only solution to the problem of war was to transform the economic system itself: from one driven by for-profit, private greed to one driven by common, public, state-run ownership of the means of economic production. Everyone should create and *contribute what they can* – and in return, everyone should *receive only what they truly need*. Take away the private profit motive, Lenin thought, and you'd take away the reason for wanting to capture and steal resources by force.[34]

You don't have to be a communist radical to agree with this general thesis about money and material gain being the real reasons behind war. Contemporary critics like Stephen Pelletiere and William Clark have argued that America's recent wars in the Middle East – the two Iraq wars, in particular – have been, above all, about securing supplies of oil and gas to satisfy the vast American consumer appetite for fuel.[35] And no less a conservative, establishment figure than Dwight Eisenhower publicly endorsed this general thesis in his famous 1961 "Farewell Address." (Eisenhower had been Supreme Commander of Allied Forces in WWII, and a two-term Republican president of the US in the 1950s.) In his much-cited address, he warned Americans about the suspect, growing influence of "the military–industrial complex." He was referring to a group of people and companies – both inside and outside of the military – with a greedy, vested interest in the business of war. He warned even further about how these companies – weapons manufacturers, in particular – donate heavily to the political campaigns of politicians who support them, creating an insidious military–industrial–congressional triangle, each side of which has its own selfish interest in big military spending, and in selling weapons around the world. You can't fight a war without weapons, right?[36] One stunning fact is this: total annual military spending world-wide is at least US$1.6 trillion, *over one-third of which* is spent by the US alone. (The US is by far the world's biggest military spender, followed distantly by China, Russia, Saudi Arabia, India, UK, France, Japan, and Germany.[37])

The upshot is that there is an elite, wealthy group of people with a selfish interest in violence and war – *the business of war* – and they are often well-connected to high-ranking officials with decision-making power in wartime. Critics of the George W. Bush administration (2001–9) called attention to the close personal connections between such high-ranking members as Dick Cheney (vice-president) and Donald Rumsfeld (defense secretary), and various oil and gas companies, weapons companies, and even PMCs.[38]

Another thing to note about the military–industrial complex is the North–South dimension of the weapons trade. This is to say that the majority of arms manufacturers are from the developed North, and they sell weapons into the underdeveloped South, enabling wars in the developing world. (The US is the world's leading arms seller, accounting for around half of all global sales, followed by Russia and France.) Critics find this exploitative, and point to instances when this has come back to haunt the selling country. An example: in the 1980s the US provided extensive weapons, military training, and guidance to Afghan rebel fighters, who were resisting the invasion and occupation of their country by America's Cold War nemesis, the USSR. A number of those "allied" Afghan fighters of the 1980s went on to become "enemy" terrorists of the 1990s and 2000s, using their weapons and training against the US itself.[39]

Perhaps a quick further comment on the connection between military spending and the economy is relevant. Experts agree that, in the short term, increased military spending *benefits* economic growth, as governments pour money into their economy with the purchase of domestic-made weapons and the like. But it's thought that, in the long term, excessive military spending *hurts* economic growth, as it forces government to grow and "crowd out" the more productive private sector. It's often been argued that, more than anything, this is probably why the USSR lost the Cold War: it went broke spending all its money on its ferocious military machine, leaving other sectors of society – basic infrastructure, schools, hospitals, scientific research – to fall apart.[40]

Another consideration in this "money and materialism" thesis harkens back to our wild, evolutionary past, and puts into question the degree to which war might be rooted in nature rather than culture. There's evidence that species of apes, chimps, and monkeys engage in war-like behavior: grouping together as a tribe, even collecting primitive weapons like sticks or rocks, getting themselves all "fired up" into a fighting frenzy, and then attacking a nearby group. Anthropologists and biologists who have studied such behavior attest that, to the degree to which such behavior has a rational explanation, it is materialistic: to expand their group's territory, to gain food, and to capture more females for breeding.[41]

1.3.5 A Feminist Analysis

Feminist thinkers like Susan Faludi have argued that men are the main cause behind war, and that, for this reason, any solution to the problem of war must involve deep, systemic, and wide-spread promotion of women into positions of power and influence around the world, even more so than what we've seen recently, for example, with Angela Merkel's success in Germany.[42]

Different feminists propose different reasons why men are behind war. Some of the most common amongst them include the following:

- *Biology and Biochemistry.* See the above about our monkey ancestors. Consider also that men have more testosterone, which has been linked to greater degrees of violence and risk-taking. Men are simply more aggressive than women. Combine that with male social power ("patriarchy": i.e., the historical dominance of men over women, across nearly every culture and country), and you have a recipe for the war-soaked social history we have suffered through as a species.[43]
- *Competitiveness.* Men are said to be more competitive than women – "alpha males," emphatically – and frequently, when their competition cannot be contained, it spills over into violent conflict. The implication is that men are worse than women at controlling anger and resolving conflicts peacefully.
- *Correlation.* This argument infers male guilt for war from a host of damning correlations. War is present throughout human history, and men have held the clear preponderance of power throughout history: coincidence? Feminists think not. And who are the main leaders during war? Men. The vast majority of soldiers? Men. The main war heroes – and war criminals? Men. The principal CEOs of weapons companies? Men. The chief scientists and engineers behind weapons development? Men. The primary audience for the seemingly endless parade of movies, books, video games, TV shows, and documentaries about war? Men. Add up all these correlations, and a vivid picture emerges of a profound male fascination with, and complicity in, warfare in all its forms. From such a degree of correlation, it's hard *not* to suppose causation.[44]
- *The assertion and expression of social power.* Feminists like Faludi argue that war is an expression of everyday, and historical, male

dominance. An even shrewder argument is that *war is itself a tool for helping ensure* male dominance over women. Faludi has argued that war is rooted in (a) a male myth about needing to protect women from other men, and (b) a male fantasy about being heralded as the mighty hero of his tribe. These two primordial psychological drives explain male propensity to go to war. Further, men have parlayed their success in war into stronger social control more broadly, and over women in particular. For instance, war heroes win political power; war profiteers make lots of money; the fortunes of war determine the fates of nations; and women and children often form the majority of civilians impacted by military activity, especially in the developing world. Taking Clausewitz's proposition that "war is the continuation of politics by other means," feminist scholars like Catherine Mackinnon opine that war is the continuation of patriarchy – both by "other means" and more directly, verging on a form of perpetual gender terrorism.[45]

1.3.6 War as an Irrational Impulse to Destroy

What all the theories above share in common is a conviction that war has a *rational* explanation, a goal-oriented causation: war is about getting power or money, or territory and resources; or it's about men, or a national culture, seeking dominance; and so on. But others have theorized that war has a more intractable dimension, one that bodes very ill for any supposed attempt to solve it. In 1920, famed psychologist Sigmund Freud wrote *Civilization and Its Discontents*. At the time, just after WWI, people wondered how such a brutal, pointless war could have followed a sustained period of peace, progress, and economic and technological growth. How, when things were going so well, could they have ended so badly?

Freud's answer was that humans, as a species, have two basic and innate drives: for *creation* and for *destruction*. (These are further linked, for Freud, to the one, omnipresent, foundational human drive: the selfish pursuit of sexual pleasure.) We are driven to create, and we are driven to destroy – and not always for good reasons. Think of a child who spends a day at the beach, building a sand castle. There's all that wonderful, careful creation – with great pride upon completion. Yet many of us have known children – or *were* such children ourselves – who, when sunset falls and it's time to go, have taken an equal joy in smashing apart their own creation in a fit of destruction. There's not much rationale to it – wouldn't it be better to leave for others to enjoy? – just a raw, primal impulse to wreck. As the child goes, so goes civilization, Freud concluded: we alternate between periods of peace, prosperity, and progress, on the one hand, and war, poverty, disease, and destruction on the other. It doesn't make much sense – there's no overall, rational story or narrative – but it's the fate of humanity. War is part of who we are: war and humankind will forever walk on together.[46]

A slightly different take on this argument comes from the anthropologists mentioned above, and even distinguished military historians like John Keegan. They argue that war is truly rooted in *a primordial, non-rational tribalism*: an instinct for, and association with, one's group or tribe which at the very least is profoundly cultural and historical – and may actually be deep-rooted in human biology, thanks to our animalistic evolution. The theory here is that it's group membership, and the drive to belong, and the desire for one's group to have primacy over others, which is the ultimate cause behind warfare. It's a variant on this vein of thinking in that there's not much rationality to it: it's just a brute fact of who we are – and very often with bad consequences that we'd be better off avoiding (as the WWI generation would attest, and I suppose we ourselves have seen a taste of, in Iraq and Afghanistan). Michael Gelven, in his book *War and Existence*, perhaps encapsulates the essence of this view better than anyone when he writes (with pithy, disturbing eloquence) "War is rooted in the existential concern for the priority of the We over the They."[47] If true, this is not a view which affords much hope in terms of liberating humanity from warfare. At the same time, does this view imply we simply do nothing in the face of such brutal violence, such costly and searing suffering? Should we therefore just lament about what psychotic, tribal, mega-violent monkeys we are, and passively accept our fate? Not many of the political theories we're now going to explore advocate such an attitude. Let's explore why.

1.4 Conclusion

Our concern in this chapter has been to lay some important groundwork in advance of our detailed investigations into realism, pacifism, and the combination of just war theory and international law/laws of armed conflict. We did this by offering an initial definition of war, analyzing its component elements. We also canvassed some basic military terms, leading to an understanding of how belligerents tend to pursue their objectives. Along the way, we learned about different kinds of weapons and tactics. Historical cases were offered throughout as helpful illustrations. Finally, we offered several explanations for why groups come to blows in warfare, and asked questions about what this says about humanity, about the state of the world in which we live, and about how we might understand, and improve upon, our unfolding future.

2
Realism: Power, Security, and Nationalism

2.1 Realism in General

Realism is a comprehensive doctrine, both of foreign policy in general and as applied to war in particular. It is so-called as it sees itself grasping the fundamental reality or truth of the situation: that politics are all about power and thus, as Clausewitz said in the last chapter, war is simply a more forceful, extreme way of pursuing power, which in general is the ability to get what one wants. Politics is here seen as an endless struggle between groups and individuals, both national and international, over who gets to control what goes on.[1] Moreover, the omnipresence of power, and the relentless, high-stakes struggle over it, renders fruitless any appeal to ethics, morality, law, and justice, whether in national or international politics, and emphatically with warfare. The one and only reality, for state and non-state actors alike, is tending strictly, even fiercely, to one's own security and self-interest. Realism, writ large, is about "Looking Out for Number One."[2]

The procedure for examining our Big Three doctrines – of realism, pacifism, and just war theory/laws of armed conflict

(JWT/LOAC) – will be, first, to describe them in detail and with respectful charity, in particular to grasp their unique insights into armed conflict. It's not for nothing, after all, that they've each survived and thrived for so long, and been so influential, for so many people. Yet we won't shy away from critical evaluation, either, and the approach in that regard will be, not so much to find the one "true," or even "best," perspective: rather, it will be more the method of "and here's what the rival theories think about that claim." *It's about putting the Big Three into intensive, illuminating interaction.* (And this debate will be spread out over the whole book and so, for example, we can't expect everything to be settled, or even raised, in this chapter.) Finally, we've seen that each of the Big Three sports internal differences and divisions – and we'll explore these, as that will also help us come to a fuller understanding of the nature, strengths, and weaknesses of each doctrine.

In the first instance, and for the sake of simplicity, we'll begin this chapter by focusing mainly on state governments and their foreign policy choices, one of which is war. This is natural both historically and also because such governments are, at least in their own territories, usually the ones who already hold the preponderance of power and are quite keen to hang on to it. Yet it's clear that adherence to (a kind of) realism is also an option for non-state actors, emphatically as applied to armed conflict, as realism is focused on winning (or, at least, not losing), on military capability which can secure one's objectives, and generally feeling permitted to do whatever one perceives as tending toward those goals. Let's develop the influential doctrine more systematically, and then conclude with a rigorous examination of realism's strengths and weaknesses.

2.2 Classical versus Structural Realism

We've seen that there's a standard and prominent initial distinction between classical realism and neo- or structural realism. I've heard the core difference between them portrayed pithily, thus: classical realists believe that "human beings suck" whereas structural realists think that "the world sucks." The difference between the two forms is akin to the difference between nature and culture. As Thomas Hobbes declares on behalf of classical realism: "The general inclination of mankind is a perpetual and restless desire of power after power that ceaseth only in death."[3] Whereas, representing structural realism, Kenneth Waltz suggests we look not within the hearts of humans but to the set-up of the international system for the ultimate cause of mass violence: "With many sovereign states, with no system of law enforceable among them, with each state judging its grievances and ambitions according to the dictates of its own reason or desire – conflict, sometimes leading to war, is bound to occur."[4]

An important term of art employed by realists is "the assurance problem." "Assurance" is a fancy word for trust. Perhaps the prime reason why smart people, and smart countries, should stick to selfishness (or group/tribal/national egoism) is that *you simply cannot trust others*. For classical realists, this is because human nature itself is, as was described at the end of the last chapter: short-sighted, short-tempered, self-absorbed, mega-violent. Only a fool would trust such creatures to be co-operative, or to care about others (or, at least, anyone outside their core "inside" group of family, friends, and "tribe"). For structural realists, countries cannot trust other countries for two major reasons: first, there are too many deep differences between them (of geography, demography, history, culture, religion, language, values, etc.); and, second, there is no effective world government to help enforce co-operative behavior that would benefit everyone, or indeed even keep the peace between them. When push comes to shove over crucial issues – like territory, resources, or pursuing or resisting hegemony – countries are fundamentally on their own. Thus, *the only smart policy is a selfish policy*. Being too "nice," too trusting, too idealistic, is a recipe for getting exploited and "suckered" in ways that may cause you serious harm.[5]

Consider what are probably the two classical cases, or sources, representing each form of realism. First, Thucydides' *Melian Dialogue* offers a frank expression of classical realism, with its ruthless overtones. This memorable, well-known piece describes a historical meeting between ancient Athenian generals and the leaders of Melos, a tiny Greek island. The expansionist Athenians want to annex Melos and supplement their power and resources, whereas the Melians wish to preserve their independence and protect their own way of life. The generals propose that "all fine talk of justice" be put aside, and that everyone stare reality square in the face. This reality is that "they that have odds of power exact as much as they can, and the weak yield to such conditions as they can get." It's rule or be ruled in our rough-and-tumble world. The Melians refuse to play along, citing their right as an independent and peaceful community not to be invaded. They subsequently get attacked and crushed by the Athenians. Both pacifism and JWT/ LOAC would deem this as ultra-aggressive and deeply objectionable, whereas classical realists simply view it as the way of the world.[6] Such realism is typically associated with such dark, pitiless, and amoral axioms as: "Divide and conquer"; "If you wish for peace, prepare for war"; "All warfare is based on deception"; "War is hell"; and, perhaps most tellingly, "All's fair in love and war."

One of the hallmark pieces of structural realism, by contrast, was penned by George Kennan at the dawn of the Cold War in the late 1940s/early 1950s. Kennan was an American diplomat, and he foresaw that Soviet Russia, in the aftermath of WWII, was going to be a potent adversary for America, as the USSR not only represented a totally different socio-economic and political system, it showed quite aggressive, expansionist tendencies during this time throughout Central and Eastern Europe especially. Did he therefore recommend bellicose confrontation with the USSR? Quite the opposite: he observed just how resourceful Soviet Russia was, how committed it was to dominating that region (especially to punish Germany for WWII, and create a buffer between itself and Germany), and the degree to which any war with the USSR would be very difficult and extremely costly for America, emphatically in a world with atomic, and then nuclear, weapons. Kennan is thus commonly credited with being one of the main architects of what is now called "containment policy": avoiding direct military confrontation and instead

patiently, and on a variety of fronts (economic, political), working
over the long haul to encircle and reduce the influence of one's
adversary, in this case Soviet Russia. And so – manifestly *unlike*
the militaristic, war-mongering Athenian generals described by
Thucydides – structural realists like Kennan have often displayed,
and argued in favor of, *a deep pragmatic caution regarding armed conflict*,
and the need to be mega-mindful of the details – facts and risks,
costs and opportunities – of each situation before deciding on what
is the smartest, selfish course of action. Analogies to chess matches,
or strategic games like "Risk," definitely come to mind. Today's
dilemmas regarding what to do about North Korea or Iran (or,
indeed, Putin's Russia) owe much to neo- or structural realists like
Kennan, and, while they may always imply *a willingness to consider* a
military option, they may actually in the end strongly recommend
against it, if the calculation doesn't pan out in one's favor.[7]

Such an approach is far too flexible and unprincipled for pacifists, of course, who have it as an *absolute rule never* to resort to
armed conflict, for reasons we'll discuss in the next chapter. It
also tends to un-nerve JWT/LOAC, especially when it comes to
the question of permissible weapons and the means and methods
of warfare: the main subject of chapter 5. There, we'll see that
JWT/LOAC levy *firm prohibitions against* the use of certain tactics,
notably: taking deliberate aim at unarmed civilians; using such
legally prohibited weapons as flame-flowers and nerve gas; and
employing such brutal tactics as torture or the use of child soldiers.
Now, it's true that realists of all stripes refuse to rule out anything in
advance, even such profoundly controversial things. They reason:
who knows how the war will unfold, and what may or may not
be useful? It all depends: in theory, anything goes. Yet in practice,
most structural realists tend to believe that very rarely, if ever, will
it actually be in one's smart, long-term self-interest to perform
such acts, or to deploy such heartless strategies. This is especially
true considering how the enemy is highly likely to retaliate in
kind, how such vicious tactics will turn the civilian population
against you, poisoning any hopes for a smooth ride after the war,
and perhaps even one's own people, and/or one's international
allies, will turn against you as well. Structural realists may, there-

fore, actually urge adherence to many of the rules of JWT/LOAC. It's just that the *reasons behind endorsement* of such principles will be totally different: reasons of morality, ethics, and justice for JWT/LOAC; and reasons of smart, long-term, self-regarding prudence for realism.[8]

2.3 More on Power, and Further Realist Insight into War's Origins

That's helpful for introducing us to the core of realism, as well as the basic contrasts with the other two Big Theories. Yet we can deepen our understanding and sharpen our analysis of realism, and then rejoin the debate, by further detailing its most frequently used concepts. Much realist talk about warfare today is developed within so-called "IR theory" ("international relations theory").[9] IR theorists, amongst other things, classify different countries based on their degree of power, usually conceived as the nexus between *hard power* (i.e., one's degree of military strength plus economic resources) and *soft power* (i.e., one's cultural and political influence around the world). Ideally, as Joseph Nye proclaims, a country most wants a "smart power" blend of both hard and soft power, maximizing one's chances for survival, security, and progressive growth in affluence, influence, and the overall well-being of one's people.[10] The IR power classifications of countries, though contested, are very influential amongst those working in foreign policy, military strategy, and international affairs, and will enable us to develop more fully the fruitful realist account of what causes war, beyond the observation in the last chapter that it must revolve around the pursuit of power, and the clash over relative power in the international system.

2.3.1 Great Powers

The great powers are those countries with the richest economies *and* the largest and most potent military capability. Their decisions often have world-wide impact. The great powers include the US, Britain, China, France, and Russia, which together happen to be the five permanent members of the Security Council of the United Nations (UN), which is nominally tasked by international law to control armed conflict, and more broadly ensure global peace and security. (More on the UN, and its Security Council, in chapters 4–6 on the JWT/LOAC.) Other countries – like Germany, India, and Japan – may come close to being great powers too, but, in terms of *both* hard and soft power, and *both* military and economic capability, the first five are the most frequently mentioned as the truly great powers.[11]

2.3.2 Middle Powers

Middle powers have a level of influence and capability in-between that of the great and small powers. Typically, they are developed, and quite wealthy, societies but they lack the population size, military force, cultural impact, and ambition to become great powers. Examples include Australia, Canada, Sweden, and South Korea. The middle powers are known for trying to be active and good "international citizens," picking projects in which they can play a globally constructive role. They like to present themselves as *peace-keepers*, for example, in contrast to the more potent military posture typically put on by the great powers. The middle powers tend to be like-minded and mutually supportive; unlike the great powers, which are often rivals and tend to butt heads on many issues (think especially today of the US, China, and Russia). The middle powers are amongst the strongest and most enthusiastic supporters of global co-operation, international law, and such international institutions as the UN.[12]

It sheds further light on realism to know how realists interpret such middle-power behavior: crucially, NOT as sincere, idealistic altruism but, rather, as *the most rational form of selfishness for countries with that kind of power-profile*. Combining together, for example, the middle powers have more influence vis-à-vis the great powers. And supporting global institutions and international law could be an important way to try to constrain, or keep a check on, the impulses of the great powers. It's sometimes said by realists that, though every country does behave like a realist, not every country pursues its self-interest in the same way on the world stage: more powerful countries tend to be "straightforward maximizers" who openly, directly, and boldly pursue what they want (they are transparently self-focused); whereas countries of lesser power are forced to be "constrained maximizers," pursuing their self-interest more through indirect and opaque strategies of "nice-ness," co-operation, and strength-in-numbers. A quick, broad contrast of the foreign policies of, say, Trump's America and Trudeau's Canada seems to bear this out.[13]

2.3.3 Regional Powers

There's a handful of countries which are quite unique in terms of power. On the one hand, they aren't plausibly described as great powers, as their influence is not truly global. Yet, they aren't classic middle powers either: usually much larger, demographically; often, not so co-operative internationally; and usually with strong militaries. They are often called "regional powers": they have many of the traits of a great power, but the scope of their influence is regional and not global. Examples include: Brazil, within the context of South America; South Africa, within sub-Saharan Africa; Turkey; and both Saudi Arabia and Iran (i.e., there can be more than one regional power within the same region, and very often they are determined opponents).[14]

2.3.4 Small Powers

Small powers can be either developed or developing, richer or poorer, and they have only a small degree of impact on international decision-making and global life. Typically, their population, economies, territory, and militaries are simply too tiny for them to exert much influence globally. Examples range from Ireland in Europe to Chile in South America, and from the Caribbean islands to those in the South Pacific.

2.3.5 Rogue Regimes (more on Realist Causes of War #1)

Recently, some countries have come to be branded – controversially – as "rogue states" (or "outlaw nations"). The term "rogue" is used to characterize actions or actors that are undisciplined, "irresponsible," or unpredictable. The idea here is of a country that is a bad international citizen. It refuses to play well with its neighbors, and actively creates trouble and instability on an international scale. It breaks the rules, goes its own way, makes the world worse, and is difficult to rein in. Most simply, rogue regimes are troublemakers.[15]

But note that the "trouble" can actually be very serious, dangerous even. For example, some state governments support terrorist groups, either by giving them weapons or safe haven. That's considered rogue state activity and, obviously, makes the world less safe. The figure of the rogue state also raises a crucial question for our ongoing meditations on the origins of armed conflict: *could regime type explain the outbreak of war?* Are certain *kinds of government* more likely than others to cause and spread armed conflict, either domestically or internationally, whether directly or indirectly? If so, then it's too pat to say that war is caused by a never-ending clash over relative degrees of power: the issue of *how power is used and exercised* may also be implicated in war's beginnings. We'll return to this below, and again in chapter 4 (in connection with authority and legitimacy, as defined by JWT/LOAC) and chapter 7 (in connection with the democratic peace thesis).

The states most frequently mentioned as potential rogues today include Iran and Saudi Arabia, as well as North Korea. Iran and Saudi Arabia are both allegedly involved in supporting some terrorist groups, and the consequences of their wide-ranging rivalry has destabilizing effects throughout the Middle East, notably Syria. Iran is also pursuing nuclear weapons, and still officially denies that Israel has the right to exist as a country. Saudi Arabia, apart from the above, has inserted itself into Yemen's ongoing civil war and also has the reputation, historically, of using its enormous oil supply to manipulate the world market price of oil to its own maximum advantage, regardless of negative effects on other countries and even the average consumer. North Korea, which already has nuclear weapons, has served for a long time as the very definition of "rogue": flouting international law and common standards of conduct, being very unpredictable, harassing its neighbor South Korea, threatening almost everyone (except for China), and selling weapons to terror groups as a needed source of revenue. We'll see if recent peace overtures alter this deep historical pattern of behavior. One final observation here as to which countries count as "rogue regimes": a number of commentators have noted how, on occasion, there actually seems very little difference between a supposed rogue and, say, a cranky great power determined to get its way, like the US invading Iraq in 2003, or Russia "annexing" the Crimea in 2014.[16]

2.3.6 Failed States (more on Realist Causes of War #2)

A failed state is one in which a government exists – there is a coherent country – but can no longer effectively govern its people or provide for their vital needs. It fails to do the most basic things people expect from their government: keep the peace, protect them from foreign invasion and malign non-state actors, enforce law and order, and ensure that essential social services (food availability, water supply, garbage collection, electricity) are provided to the people. Based on these criteria, Somalia, Syria, and a number of countries in sub-Saharan Africa are failed states. Perhaps Afghanistan, Yemen, and parts of Iraq would count, too.[17]

Failed states are often produced by wars, and they often cause wars, too, as groups fight over the state collapsing before their eyes. And so, another important variant on the basic realist account of armed conflict is that the *absence of power*, or a *power-vacuum*, also might provoke chaos and violence and create a downward spiral into war. For, failed states produce massive refugee flows across borders, as people, struggling to escape famine, poverty, or bloodshed, flee the failed state for other, safer countries. Failed states can create civil wars, as just noted, and/or they can become havens for terrorists and criminals (consider the waters off the Somali coast, notoriously aswarm with pirates eager to hijack and loot foreign vessels). Such states can, at their extreme, become plausible targets for armed humanitarian intervention (AHI), which is to say the use of military means to enter a foreign country for the sake of providing humanitarian supplies and/or ending a grievous threat to the lives of the people within that country. The US attempted to do so in Somalia in 1993, in a half-hearted way which was not successful. This was probably a major reason why it failed to intervene during the genocide in Rwanda the following year, wherein more than 800,000 people were killed. The failure to prevent such slaughter spurred the development of the so-called Responsibility to Protect (R2P) doctrine.[18]

We'll discuss AHI, and cases like Somalia and Rwanda, in greater detail in chapters 4–6. We note now, though, how such cases perhaps reveal one of the clearest differences between realism and JWT. A case like Rwanda is often offered up as a paradigm instance where, from the perspective of morality and justice, a timely military intervention ought to have happened, to protect the Tutsi people and moderate Hutus from the murderous designs of the Hutu extremists. Realists, by contrast, while perhaps not denying the moral tragedy of the situation, tend nevertheless to be very skeptical and critical about using military power for any reason other than a strict calculation of national self-interest. If no such national interests are involved (as they weren't, they'd say, in the case of Rwanda from the perspective of US or other Western interests), then non-involvement is what realists recommend, to the dismay and ethical criticism of just war theorists.[19]

2.4 The Tools of Foreign Policy

War is but one tool of foreign policy, and it's important to briefly note them all, at least as standardly understood. *They are relevant to each of our three major doctrines on war moving forward*, as: (a) realists want to know, and make use of, every tool they can to maximize national self-interest; (b) pacifists want to know all the tools, and creatively try to invent further ones, instead of using the one "final," and for them prohibited, tool of warfare; and (c) throughout JWT/LOAC, there is commitment to the use of force "only as a last resort," after all other plausible means have been exhausted. So, we should canvass what those are. If you crack open a textbook on foreign policy and IR theory, it'll inform you that, fundamentally, there are "four legs to the table," or four standard "tools in the tool-box" for countries to use to try to maximize their interests internationally: diplomacy; positive economic incentives; negative economic incentives; and armed force.[20]

2.4.1 Diplomacy

Diplomacy is the attempt to persuade another country, either through praise or criticism, through rational argument or emotional manipulation, to adopt your view and act accordingly.[21]

2.4.2 Economic Incentives

Economic incentives involve the use of money as either a carrot (positive incentive) or a stick (negative incentive) in an effort to gain leverage on another country and influence them to do what you want. A "carrot" may be a thinly disguised bribe, or a mutually profitable trade deal. "Sticks" may be wielded by imposing fines on a country, slapping taxes or "tariffs" on goods which that country exports into your own, or, at the extreme, imposing economic sanctions.[22]

2.4.3 Sanctions

Sanctions are deliberate actions that you believe will thwart the interests of the other ("target") country. Sanctions can vary in level, intensity, and effect. Small sanctions include mostly symbolic actions like withdrawing your ambassador from the target country, or expelling the target country's diplomats from your own country. You are clearly conveying displeasure, and punishing that society, but only to a tiny extent and in a way very targeted toward the government of that society.[23]

What about economic sanctions, specifically? Here we must distinguish between *targeted* and *sweeping* sanctions. Targeted sanctions are measures of punishment, non-cooperation, and interest-thwarting aimed at hurting only the elite decision-makers in the target country. An example related to war occurred in 1991, when the US slapped targeted sanctions on the leaders of a military coup in Haiti. (A *coup d'état* is a sudden, usually violent, displacement of the current government by a small group of plotters who now assume leadership.) The Haitian coup leaders had their American bank accounts and assets frozen, which is to say that the leaders could not access or sell them. The coup leaders, and their families, were forbidden from entering the US and, perhaps most crucially, the US was able to use its influence around the world to get many other countries to do the same thing. After a few months of such pressure, the Haitian coup plotters gave up power. The goal of such measures is clearly to punish, but *to be discriminate in whom one is punishing.* The punishing country is leaving the common people out of it, concentrating instead on "sticking it" to the elites. Other subjects of targeted sanctions today include Iran and North Korea, for alleged "rogue" behavior. Syria has had targeted sanctions put on it, to try to damp down the civil war; and Putin's Russia has, too, as the West has sought to punish its military moves regarding Crimea and Ukraine in particular.[24]

Sweeping sanctions are measures of punishment and non-cooperation that either deliberately target, or at least directly affect, the majority of citizens in the target country. A recent example is America's longstanding embargo on trade with Cuba, to oppose Cuban communism and the old regime of Fidel Castro. It began in 1962 and continues to this day, although there have been recent relaxations.[25] Another recent example deals with sweeping sanctions leveled by the international community – particularly the US and the UK – at Iraq following its invasion of Kuwait in 1990. These sanctions were largely maintained after the end of the Persian Gulf War in 1991, on grounds they were needed to ensure Iraq's full and continuing compliance with the peace terms of that first war, notably the weapons inspection process. During the 1990s the sanctions caused the erosion of many important quality-of-life factors within Iraq: the economy shrank; infant mortality substantially increased; rates of easily preventable illness substantially increased; life expectancy decreased; unemployment rates shot up; illiteracy increased; and so on. As a result of the severe and sweeping effects of these sanctions, the US lost much international goodwill and co-operation in its dealings with Iraq, notably in 2003, when the US decided to invade that country.[26]

Countries sometimes view sweeping sanctions favorably, usually when relations with the target country are hostile and sanctions seem like a more moderate, non-violent alternative to war. Some pacifists, for example, commend sanctions over war. But there is little historical evidence that sanctions work in changing the offending policy of the target country. For example, Castro remained in power for decades. And the sanctions on Iraq didn't affect Saddam Hussein's grip on power or his attitude toward weapons inspection. And they often disproportionately hurt innocent civilians – a fact that seems largely responsible for what appears to be, today, the negative attitude toward them.[27]

2.4.4 Armed Force

When relations have completely collapsed, and perhaps when the other three traditional tools of foreign policy have failed, countries sometimes resort to violent force in their relations with each other. This can be small-scale or pinpoint force (or even just "show of force," like a military drill or "war games" exercise, close to the target's borders), or it can be a classic, out-and-out, direct, "hot," shooting war as defined in the last chapter. This book mainly deals with this final option, of course, and so that's all we need say here, save to note that such makes for a nice transition back into a more targeted analysis and evaluation of realism's attitude about war, especially from a pacifist or JWT/LOAC perspective.

2.5 Rival Criticisms and Realist Responses

2.5.1 *Enduring Principles versus Flexible Calculations*

We saw above, in section 2.2, how there may actually be some potential overlap, and mutual endorsement of principles, amongst our Big Three theories: structural realists like Kennan may side with pacifists against JWT/LOAC in terms of refusing to go to war, for instance in the Cold War case, where justice might have demanded armed resistance against Soviet tyranny yet sober calculation of the probable costs of doing so advised strongly against it. And realists, of either the classic or structural variety, might actually be willing to adhere to JWT/LOAC rules and restraints on the methods of war-fighting, again if the long-term self-interested calculations favor such.

That said, it must be noted where enduring differences, and criticisms, lay. We know that, as a matter of principle, pacifism demands an *absolute* prohibition on war, which realism can't agree to. And we know that realism is only going to agree to JWT/LOAC-style rules and regulations regarding warfare if it looks like such rules are going to be in one's self-interest. If not (and some suggest that adhering to JWT/LOAC rules would be like "fighting with one arm tied behind your back"), then realism would recommend *ignoring the rules and doing what is needed to win*. That's not something which JWT/LOAC can endorse, as the principles are thought to have abiding moral resonance, and not merely rest on contingent, self-interested calculation. There are a great many historical cases, just war theorists would point out, of belligerents getting those calculations wrong, leading them to make dreadful mistakes harming not only others but even themselves: many a would-be conqueror (Hitler, Tojo, Saddam) has come to considerable, even lethal, grief by supposing that things were going to turn out far better than they actually did. Moral principles don't change, pacifists and just war theorists would say, whereas the calculations of self-interested cost–benefit do so constantly, and moreover are subject to all kinds of bias, rash emotionality (especially fear, anger, and pride), imperfect information, failure to understand the opponent, and very frequently a short-term over long-term perspective. Wouldn't we just be better off adhering to firm principles which never change?[28]

Realists wouldn't back down, though, as they view their flexibility as a huge virtue and not a vice. It's a complex, rough-and-tumble world, they'd say, and countries would be foolish to take any option off the table for the sake of principle. After all, unscrupulous opponents, knowing such principles, *may well use such scruples against you*: for example, by locating a militarily valuable target within a civilian population area, knowing that you won't attack it and it'll thereby be preserved. Or, worse, using child soldiers against you, knowing that you'd be extremely reluctant to strike at them. Everything ought to at least be considered. As a result, so much of realist writing and thinking about war, and war tactics, tends to revolve around the factual things detailed earlier in this chapter, and the last, such as: What are our military capabilities and weapons? What kind of power are we, and what kind of power is our opponent? What suits us better, and them worse? Which mix of tools in the foreign policy tool-box should we use? Let's gather as much information about them as possible (as "knowledge is power," after all – a favorite realist maxim), and hone our own tools and capabilities to the greatest degree. Such factual preparedness will, it is hoped, cut down, as much as humanly possible, on the concerns above regarding how constant, variable, flexible calculations can sometimes go seriously astray.[29]

2.5.2 *What if Everyone Were a Realist?*

Pacifists in particular tend to argue that, if everyone were a realist, the world would be a horrible place indeed: bleak, uninspired, paranoid, dangerous. We'd all be caught up in a never-ending game of strategic selfishness which would be violent, exhausting, dreary, and dreadful. It would, the pacifists say, hardly be a life worth living: we'd be real-life prisoners, to speak, caught perpetually in our own dilemma.[30] Indeed, it's been said that the recommendations of realism, if widely adhered to, would actually make for a self-fulfilling prophecy: in urging everyone to behave in a totally selfish, un-trusting, defensive, power-obsessed kind of way, realists actually make it more likely that that's indeed what will happen, resulting in a sub-optimal, insecure, shabby world. And then that "fact" encourages *even more* such behavior, resulting in a grim downward spiral. Amidst such bleakness, who'd have the courage to be optimistic? Who'd gamble on co-operating or trusting (much less, loving) someone, or on creating something new and promising of benefit to us all? How, as the utopian William Morris asked, would the world ever become better if everyone believed that it sucks, and that human beings suck besides?[31]

Realists couldn't disagree more. They argue that, if everyone were a realist, the world would actually be better, at least in the sense of being more secure. When people are too trusting or too idealistic, they make mistakes and can open themselves to being exploited and used. Realists pride themselves on how factual and detail-oriented, how situation-specific, they are, and view it as a serious weakness with pacifism and even JWT/LOAC how they resist such situatedness in the name of trying to impose universal principles from above. Emphatically when it comes to war-fighting, realists argue that theirs is, at least, *the safest option*: you are the least likely to make serious mistakes that will endanger your people if you tend very firmly to the maximization of your own national self-interest. This may mean the world can't be as great as it might be if everyone were altruistic optimists but that, they'd swear, is just so far afield when it comes to armed force and international relations that it's not even worth spending much time thinking about. *The world is not your business when you're at war:* only the survival and safety of your own people.[32]

2.5.3 Nation versus World

This relates to a foundational difference in world-view. Realists – even sophisticated ones like Kennan – argue that a national government is the representative of its own people, and owes to them not merely its first duty, but really *its only substantial duty*, and that cross-border, cosmopolitan concern is not at all required in a such a fractured, tribal world as ours. National communities are moral communities proper, and they root things like rights and obligations. Duties between countries, or to distant strangers, are at most purely voluntary, and subject to the all-pervasive calculus of pragmatic, national self-interest (as we saw above regarding Rwanda).[33] In reply, pacifists and JWT/LOAC argue in favor of non-voluntary, truly cosmopolitan duties. We'll see more fully why (and to what extent) within their respective chapters, but for now we note that they'd reply to the realists by saying things like: "National governments may indeed be the representatives of their own people, but that doesn't mean they are utterly duty-free, and may do whatever they want internationally, including at war." Consider, by analogy, our common view that, while a lawyer is indeed the agent of his client, he's not entitled to literally do *anything* he sees fit on behalf of his client (like tamper with evidence, or murder a damaging witness). Or consider, as Thomas Pogge notes, that if any group seems to be the foundation of moral connection between people (and the source of much of their personal identity), it's probably the family. But we don't usually think that people are entitled to do anything and everything on behalf of their family, like steal from others or attack the neighbors next door because you disagree with their values. So, why should realists believe that all moral duties stop at the national or tribal border, such that *anything* may be permitted in warfare, if only the calculations turn out right? Indeed, for all realist talk about the value of national community, and its strong connection to personal identity and meaning in life, others may retort that *nationalism has actually been one of the most potent and dangerous causes of war*, both civil and international, in modern times. How is that consistent with realist prudence?[34]

2.5.4 Biased Against the Weak?

Realism is said, by critics like Jack Donnelly, to favor and legitimize the actions of the most powerful states in the interstate system. If states ought to do what's in their best interests, and if the most powerful states have the greatest capability to do what they want, then the result is a legitimation of the fact that powerful states are able to exert such influence over the interstate system in general and smaller states in particular. But this bias in favor of the powerful is at odds with some of our other values, e.g. democracy, and the self-determination of political communities. Also, if the only standard for praise and blame in the realm of foreign policy is a prudential one, this all-too-conveniently deprives smaller, less influential states of what is often their most powerful claim for global reform: moral criticism. In other words, realism is a doctrine that powerful states – from ancient Athens to contemporary America – advance to render themselves immune from criticism. But we all know that, often, it's precisely the most powerful who deserve the most scrutiny.[35]

The realists who reply to this forceful accusation prefer to frame the issue differently: not so much "bias against the weak" (which can indeed seem cruel and discriminatory) but, rather, insisting that our chaotic world's best hope for peace and stability comes through order, and usually that order must be imposed by powerful countries, as they have the greatest capability. An old axiom of realism is "balance of power," which was the notion – popular in Europe during imperial/colonial times – that peace can actually be stably maintained if no one country becomes too powerful, and if each of the great powers takes responsibility for peace and security (and its interests, of course) within its own "neck of the woods." Critics might still find that oppressive and controlling, but realists may reply that, historically, there's some evidence that it has worked (e.g., in Europe, post-Napoleon from about 1815 to 1914, or perhaps even during the Cold War, 1945–90), whereas when there's a power vacuum (say, in today's Somalia, or sub-Saharan Africa more broadly) or one country becomes way too powerful way too quickly (Revolutionary France, 1790–1815, or Japan in the Asia-Pacific during the 1930s), then war and disruption follow. One of our era's most prominent realists, Henry Kissinger, is still very much a fan of the balance-of-power concept, suggesting such things as that, so long as such great powers as America, China, Russia, and the European Union (EU) each stay in their own historical "sphere of influence," this probably stabilizes the world – *peace through order, not through justice* – whereas when such powers interfere with each other, that's when large-scale armed conflict erupts. (If someone were to query where smaller countries fit into this schema, Kissinger would probably reply with a shrug.[36])

2.5.5 Necessity (including Military Necessity) versus Free Moral Choice

One final critical exchange for this chapter concerns another axis for understanding realism. There's a contrast between *descriptive* or factual realism and *prescriptive* or normative realism. (This distinction doesn't logically have to line up with the more prominent one between classical and structural realism.) Descriptive realism is the doctrine according to which states *are in fact* motivated by self-regarding considerations of power, security, and national interest, and *not at all* by those of morality or justice. States simply don't care about morality and justice; they only care about their own interests in maximizing their hard and soft power. They might – and often do – pay lip-service to the ethical ideals discussed by pacifism, or by JWT/LOAC, but, in the end, states always act to benefit themselves. Prescriptive realism, by contrast, is the doctrine that states *should* only so act: perhaps most plausibly on grounds of self-protection rather than a militaristic glorification of hegemony. For most of this chapter, we've presupposed prescriptive realism, with its flexible *principles*, its views about the *duties* of governments, and its placing *value* on peace through order and security. But before we go, we should pause to consider the purely descriptive side of realism, especially in its strong or extreme form, as it raises interesting questions about necessity versus free choice, and allows us to introduce the concept of "military necessity" in connection with war.

Some descriptive realists, like Edmund Wilson, argue that the international sphere is not actually a place wherein free choice – including morally or legally responsible choice – plays a meaningful role at all. These more extreme realists frequently speak of the "necessity" of state action in the global context: states have no meaningful choice but to act on the basis of power and interest *if they are to survive at all*, much less thrive. This could either be

because of the utter bleakness of human nature, or the sheer feroc-
ity of the international arena. On this understanding, war is seen
as an entirely predictable, even inescapable, reality of the interstate
system; it's a simple fact that the clash of national self-interest will
sometimes spill over into armed conflict. It's another simple fact
that states will do whatever they can, in the midst of war, to try to
win. To hope otherwise – as JWT/LOAC does – is to engage in
wishful thinking about the brutal facts on the ground in the international domain. So, attempting to limit or constrain warfare – for
example, with a set of rules and laws – is, as Clausewitz contended,
at odds with its very nature and the very animating force behind
state action. Wars are rather like storms that just naturally break
out every now and then. No one can influence them and they
just have to be left alone to exhaust themselves, as it were, and
blow over. US Civil War General W.T. Sherman once said, for
example, that "[w]ar is cruelty, and you cannot refine it ... You
might as well appeal against the thunderstorm as against these terrible hardships of war."[37]

Michael Walzer, the dean of living just war theorists, finds such
talk to be quite misleading and ill-founded. As Walzer says, rarely
is a state or community credibly threatened with extinction; so,
the day-to-day reality of international affairs is *much more a matter
of probability and risk than it is of strict necessity* and of adhering to
the fierce requirements of survival. States, it seems, are much freer
to choose between alternative courses of action than descriptive
realists contend. This means that states are free, in a very clear
sense, to choose to act on the basis of moral commitments and
conceptions of justice, as well as upon considerations of their own
national interest.[38]

It seems straightforwardly true that states do, indeed, act upon
the basis of both national interests *and* moral commitments. This
is perhaps clearest in democratic, representative regimes where the
people simply insist that their governments take seriously the moral
beliefs of the nation. Perhaps these states don't always listen, but
they do so at their own peril, since the discipline of the ballot box
remains. It follows from these thoughts that the ancient Athenian
generals, for example, were not actually forced to "rule or be

ruled"; rather, they carried out a deliberate policy of aggressive expansion that was authorized by the Athenian assembly. The crushing of the Melians was not a "necessity of nature" – an "ingurgitation" of a small state by a big one, in Wilson's biologic words – but rather the brutal outcome of a free collective decision by a group of Athenians giddy with power and lusting for more. (Walzer notes that Thucydides omits these prior debates and decisions in his *Dialogue*.) Walzer then persuasively concludes that war is the inevitable product *neither* of the structure of nature *nor* of the international system; rather, war is "a human action, purposive and premeditated, for whose effects someone is responsible."[39] Pacifists, of course, would agree.

Though Walzer admirably calls into question extreme descriptive realism as it applies to international relations overall, and to the outbreak of war in general, even he himself – as we'll see in detail in chapter 5 – does refer to a kind of necessity *in the midst of* some of the most dangerous circumstances of armed conflict. He invents something called "the supreme emergency exemption" to deal with these sorts of dire situation. It refers, essentially, to a kind of right to set aside the rules of JWT/LOAC if one's community is genuinely, imminently threatened with utter destruction or enslavement, as in rare (yet sometimes real) cases of attempted genocide.[40] We'll rejoin the issue at that time, but for now a few comments. First, the supreme emergency exemption is a piece of JWT, and *not at all* of the LOAC, which stipulates that its rules are to be followed, period. While "military necessity" does get mentioned within the LOAC, there its most common meaning is "what one side genuinely needs to do to pursue military victory – *but within the confines of the rules as already established.*" In other words, most pieces of LOAC forcefully, even stridently, declare that the laws of war have already been framed with military necessity in mind, and thus belligerents may not appeal to such a concept in order to violate or set aside such laws, like those banning certain weapons.[41]

Yet the issue is not so straightforward and, as we'll see – especially when considering the actions of soldiers – there may be desperate circumstances within war of severe duress, pressure, chaos, confusion, threat, and coercion wherein easy assertions about "free choice, and moral and legal responsibility" may well be far too easy and inaccurate. Such situations are relevant for war crimes trials after war (chapter 6), and the debate about the so-called "moral equality of soldiers" (chapter 5). Might some soldiers be excused – i.e., not punished – for some controversial actions, provided they acted truly under the pressure of necessity in the moment? Moreover, *how exactly should we understand necessity?* It can be important: one often hears, at the macro level, ordinary people using sweeping talk of "wars of necessity versus wars of choice." A prominent formula, at least to put on the table for now, is Daniel Webster's from the so-called *Caroline* Affair.

In 1837, Canada was still a colony of the UK, which itself still had tense relations with the US, following the Revolutionary War and the War of 1812. In 1837, some Canadians tried to revolt against the British Empire, and they (and some of their American friends) sailed a ship called the *Caroline* toward an island in the middle of the Niagara River, straddling Ontario and New York State. From there, they planned to stage their revolution. The British boarded the *Caroline*, killed an American in the battle, set the *Caroline* on fire and sent it plunging over Niagara Falls in a dramatic crushing of the incipient "rebellion." The Americans became outraged at the incident, doubly so when the British claimed they acted in "necessity, out of self-defense." The American senator Webster articulated a sharp, angry denial of any such rhetorical, transparently self-serving claim, asserting that true necessity would be "instant, overwhelming, and leaving no choice of means and no moment for deliberation."[42] Such a conceptual construction is still today consulted in debates, both about self-defense and prevention (which we'll canvass in chapter 4) and necessity and excuse (to which we'll return in chapter 5). For now, we note how Walzer and JWT might have the upper hand on the descriptive realists when it comes to the (macro-) decision to go to war in general, but the realists may have lingering strengths when it comes to (micro-level) experiences on the ground in the chaotic midst of a frenetic, bloody, and brutal battle.

2.6 Conclusion

Realism is one of the most influential, and time-honored, systematic doctrines about warfare and what to do about it. It views warfare from the prism of national self-interest, and as a result is obsessed with things like power, security, order, and national belonging. We canvassed how international power gets analyzed by realists, ranging from great powers to failed states to non-state actors trying to carve some space for themselves. We raised the issue of a connection between type of government or regime and the outbreak of armed conflict, showing how the realist analysis has a fruitful, manifold perspective on war's ultimate origins. We then mentioned the tools of foreign policy, as they are what countries use to forward their interests, and one such tool is war. We turned, lastly, to examine the realist attitude toward war in a rigorous five-fold fashion, serving up its pros and cons. Next up, for the same treatment, is the doctrine at the other, more idealistic, altruistic, and cosmopolitan extreme: pacifism.

3

Pacifism: Ethics, Cosmopolitanism, and Non-violence

Pacifism is a form of idealism. If realism recommends group self-ishness, lack of trust, and national egoism, then idealism – as the opposing "big-picture" perspective on foreign policy – recommends a kind of other-regarding altruism, stressing an ethical imperative not merely to forward the relative standing of one's "in-group" but, rather, *to use one's resources to do what one can to make the world a better place*. To this extent, pacifism is both individual and cosmopolitan, in contrast to realism's tribal nationalism. Individual, in viewing all human beings with ethical concern; and cosmopolitan, in denying that national borders have deep moral significance. What matters morally is that we do the right thing, and we respect everyone's lives and rights. And it's difficult to square that principle with violent armed conflict, whether for reasons of smart strategy (realism) or of justice (just war theory).

In contrast to the gritty realist ethos of "national security in an insecure world," pacifism offers a vision of "human security in a better world," where human security refers to the well-being of individual people and their feeling not merely physically safe from things like crime, terrorism, and foreign invasion but more broadly empowered and enabled to live their lives in a satisfying way.[1] A pacifist, for instance, would consider as sheer waste the enormous military expenditures by – and vast martial organizations within and across – nation-states, and experience a sad sense of lost possibility and an angry conviction regarding this wastage in comparison to what such resources could do in terms of food, housing, education, technology, health care, and research and development for the future.[2]

3.1 Defining Pacifism

There's pluralism within pacifism, as with the other traditions. Two big distinctions stand out. The first is between pacifism as non-violence versus pacifism as (mere) opposition to war. The second is between religious versus secular (i.e., non-religious) forms of the doctrine. These hook into further sub-distinctions, as we'll see. Let's consider both, explaining how, for our purposes, it's most germane to understand pacifism as opposition to warfare, rooted in secular moral reasons.

3.1.1 Non-violence versus Anti-war

To define pacifism as mandating non-violence is a distinguished linkage, and often goes together with religious forms of the doctrine. It's important to see that it's *a much more demanding doctrine* than what Jenny Teichman defines as the logical core (or common denominator) of all forms of pacifism: "anti-war-ism."[3] For, warfare is merely one form of violence, albeit arguably the most costly and destructive as well as the most historically crucial. Thus, non-violence makes further demands than mere opposition to warfare: it implies no interpersonal violence on a sub-war scale, which in general may be wonderful but, strictly speaking, may actually call into question things like the police's use of force to thwart and arrest. Such may, in the end, be hard to support, especially in the face of some ruthless criminals or terrorists threatening, or perpetrating, harm against defenseless people needing protection.

How are the police to stop such criminals while literally using no force or violence? Often – to say the least – such figures cannot be "talked out" of their malign intentions and nasty deeds and, as soon as one allows even for things like physical restraint, forcing handcuffs onto people, pepper spray, and then certainly when escalating from there to hitting with a weapon or even shooting, you are obviously allowing for violence. Non-violence may also call into question certain sporting activities, as well as the killing of animals for food purposes. Indeed, the definition of violence, on this understanding, becomes crucial, and may have far-reaching implications indeed. For example: is only *physical* violence in view, or would something like verbal abuse, "cyber-bullying," or *psychological* cruelty (or even mere attempts at manipulation), count as violence as well? Violence revolves around the intention to harm, and harm can be a concept which, while "clear in the middle," so to speak, may yet be fuzzy, vague, and contested around the edges.[4]

Moving from such human concerns of physical/mental/social/ institutional violence out toward animals, as mentioned, and perhaps even beyond that: might forms of intentional damage to other life-forms, like insects, or even to the natural environment in general, be considered violence which we must never engage in? There is, accordingly, a robust literature trying to define violence, and it must concern itself with such things, as opposition to it is the essence of this more demanding conception of pacifism.[5] And there may be problems of self-reference, as when we talk about non-violent resistance, a favored tactic of many pacifists. It's perhaps most explicitly mentioned through Mahatma Gandhi's idea of *Satyagraha*, usually translated as "soul force": his encompassing world-view regarding determined and systematic non-cooperation in the face of rights-violation, leading to positive change in the world (such as helping to push Imperial Britain out of his India, eventually succeeding in about 1947). *Where is the limit between "forceful resistance" and violence?* The former can't involve overt forms of physical violence or killing, of course, but sometimes people talk about blowing up railway bridges as part of, say, "pacifist" resistance to the Nazis in Scandinavia during World War II.[6]

While it's resistance, and laudable resistance at that, I'm not sure it's not violent. Indeed, isn't resistance to Nazis, or other aggressors and rights violators, *intended to harm them* — at least insofar as one is attempting to dismantle their capacity for brutality? Any attempt to thwart Nazis would, from the Nazi point of view, count as intending to harm them and their cause. Even just "getting in the way," by putting one's body in-between Nazi tanks (or Imperial forces, or racist police) and innocent civilians, may arguably count as intentionally trying to harm, by putting a material difficulty in the way of one's opponent, leading to queries regarding whether it's possible to avoid all forms of "violence" when one "resists."[7]

Without denying that such are important, challenging philosophical debates regarding non-violence (or *Ahimsa*)[8] as a strict and sweeping doctrine, for our purposes in this book on war it makes the most sense to define pacifism in the first instance as anti-war-ism. Such avoids the issue of self-reference, for example: one can with logical consistency be in favor of certain kinds of resistance, perhaps even violent resistance, just not *war-scale* resistance. One can simply say that it's *the kind and magnitude and destructiveness* of the violence that war involves which is forever objectionable. And it *is* what all pacifists share in common: no matter what kind of pacifist you are, you must believe that war is always wrong; there's always some better approach to the problem than warfare. The other "tools in the tool-box" of foreign policy — diplomacy, economic incentives, sanctions — are much more appealing, and defensible, than armed force. So, unlike realists, pacifists believe that it *is* possible and meaningful to apply moral judgment to international affairs. In this, they agree with just war theory and the laws of armed conflict (JWT/LOAC). But they disagree with just war theorists regarding the application of moral judgment to warfare. Just war theorists say war is *sometimes* morally permissible, whereas pacifists say war is *never* morally permissible.

It's worth mentioning that Jim Sterba has attempted to mesh JWT and pacifism into what he calls "just war pacifism,"[9] and he has collected some support in this regard, most notably in recent years from Larry May, who prefers the term "contingent pacifism." The idea of the "just war pacifist" is that one should believe in and adhere to pacifism unless and until, under (very rare) real-world conditions, a particular war might actually fulfill the exacting standards of JWT/LOAC, as we'll specify in chapters 4–6. One's pacifism is merely contingent on the circumstances of the real world: it's not theoretically absolute.[10] I appreciate these well-intentioned attempts to integrate JWT and pacifism, and such possibilities will be explored repeatedly in this volume. But I myself follow Michael Neu in disbelieving in the curious figure of the "just war pacifist."[11] It's a strange kind of pacifist, after all, who can morally endorse warfare. If they do endorse warfare under just war conditions, then I suggest that ... they are just war theorists! Sterba's proposal doesn't integrate, so much as assimilate, pacifism into JWT, thereby depriving us of a robust, time-honored, and important *independent* doctrine about war ethics. May, in *Contingent Pacifism*, seems to *derive* pacifism *from* JWT.[12] It seems fair to say that some pacifists might be nonplussed, even offended, at such a suggestion. They hold the value of peace in an absolute, and not merely contingent, sense.

The conceptual kernel of both theories, in my view, cannot be reconciled – JWT says warfare *can* be permissible, and then defines those conditions, whereas pacifism says, *regardless* of the conditions, war *cannot* be morally justified. I guess I'm saying that, for our purposes here in this book, *absolute pacifism* – and not contingent pacifism – *is what we mainly have in mind*, if only for reasons of robust and insightful conceptual contrast. Not to deny contingent pacifists the right to label themselves how they wish: just to focus on the kind of pacifism which provides the meatiest theoretical contrast and debate with both realism and JWT/LOAC. Hence: pacifism's logical core is understood as anti-war-ism. War is always wrong.

Peace is itself another contested concept, even amongst pacifists. A foundational distinction is drawn between *negative* and *positive* peace. Negative peace is the absence of war, or war-scale violence, and thus is precisely "anti-war-ism." Though undeniably the logical core of the doctrine, pacifists like Michael Fox object that this unflatteringly depicts pacifism as *merely reacting against* something – violence, a particular war, the military–industrial complex — as opposed to proposing something substantial and creative.[13] And the latter is precisely positive peace: not merely the absence of war but, more robustly, the construction of a just and secure social order wherein the absence of war endures. As Andrew Fiala says: "[P]ositive peace encompasses cooperative, tranquil, and harmonious relations and the broader concerns of flourishing and integration."[14]

The distinction and debate between negative and positive peace, which some phrase as the difference between *minimal* and *maximal* peace, is prominent within pacifist literature. There are, in particular, many rival characterizations of what is needed, and desirable, for positive peace.[15] These need not belabor us here, as they're not central to the clash with realism and JWT/LOAC. For that, for reasons explained above, we focus on a conception of pacifism which is *negative and minimal yet absolute*. We may still query whether such a conception should be religious or secular.

3.1.2 Religious versus Secular

Of course, many pacifists are so for religious reasons. Christians and Buddhists, in particular, have vibrant pacifist sects, such as the Quakers. They base their beliefs on sacred scriptures and a drive to be more like The Divine, whom they view as essentially peaceful. Buddha is credited, for example, with saying: "It is only when you find peace within yourself that you can live at peace with others." He was a wealthy prince who, when touring outside his privileged palace one day, became so distraught at the suffering of ordinary people that he resolved to renounce his pampered life and devote himself to the alleviation of suffering in others, alongside the cultivation of enlightenment and compassion. The Buddhists have an admirable tradition of deploying techniques of mental and emotional self-discipline designed, amongst other things, to ensure one never gets so angry as to lash out in physical violence with the intent to harm others.[16]

Christians, of course, point to Jesus' life and teachings as disclosed in the New Testament. These do indeed seem to recommend pacifism. Consider the imperative to love even your enemies, and the famous command: "But I tell you not to resist an evil person. Whoever slaps you on your right cheek, turn the other to him also."[17] Jesus further says that he is "gentle and humble in heart," and he never uses violence to convert others to his beliefs, and he refuses to use force to resist his arrest, commending his disciples: "Put your sword back into its place; for all who take the sword will perish by the sword."[18] Indeed: "Blessed are the peace-makers."[19] Many Christians refer to Jesus as "The Prince of Peace," and there are many references to peace in nearly every form of Christian religious worship.[20] Consider, perhaps above all, The Golden Rule itself, wherein – since I myself don't want to be subjected to violence or being killed – it seems to follow that I should never engage in violence, killing, or war myself. And it's not only the New Testament. Consider the memorable remark from the Old Testament: "He shall judge between many nations, and rebuke strong nations afar off; they shall beat their swords into plough-shares, and their spears into pruning hooks; nations shall not lift up sword against nation, neither shall they learn war anymore."[21]

But this, like much in religion, can be contested. For example, as we'll see over the next two chapters, JWT was long associated, historically, with the Catholic Church – so, clearly, the Church doesn't think that pacifism is mandated by the New Testament. (Consider Paul's proposition in Romans 13:4 that a good governing leader "beareth not the sword in vain; for he is a minister of God, a revenger to execute wrath upon him that doeth evil.") Thomas Aquinas, whose *Summa Theologica* is the basis for the official doctrine of the Catholic Church, endorsed JWT and, in fact, contributed to its core concepts, notably proportionality in the *jus ad bellum* and the so-called Doctrine of Double Effect in the *jus in bello*. (Full definitions to follow in those chapters 4 and 5.) And Augustine, inventor of the *jus ad bellum* condition of right intention, was himself an actual bishop of the early Church, and his writings became powerful inspiration for Protestant theologies centuries later, and so, it would seem that even non-Catholic branches of Christianity need not logically commit to pacifism, either.[22] Consider further that there are many instances in the Old Testament (such as *Deuteronomy* 20, or *Joshua*) where God not merely permits but fully commands the Israelites to go to war, for example against the Canaanites. And so, in a move similar to that above regarding "non-violence versus anti-war," it seems *best to opt for lesser controversy and difficulty*. Thus, I will not here delve further into religious justifications for pacifism because they rest on beliefs – about God and Scripture, souls, and the afterlife – which simply are too personal, speculative, contentious, and even exclusionary. I direct the reader to the note for more on the religious perspective.[23]

This book will stick to secular justifications for pacifism, which try to appeal to *any* rational person *regardless* of religious affiliation. The most relevant pro-pacifist arguments here include the following: (1) a more *"teleological"* form of pacifism (or TP), which asserts that war and killing are at odds with human excellence, virtue, and flourishing; (2) a more *"consequentialist"* form of pacifism (or CP), which maintains that the benefits accruing from war can never outweigh the costs of fighting it; and (3) a more *"deontological"* form of pacifism (or DP), which contends that the very activity of war is intrinsically unjust, since it violates foremost duties of morality and justice, such as not killing other human beings. Perhaps most common and compelling amongst contemporary secular pacifists – ranging from Robert Holmes and Richard Norman to Andrew Fiala and Cheyney Ryan – is a mixed doctrine which combines, in some way, all three.[24]

3.1.3 Clean Hands?

Before describing in detail – and considering critically – TP, CP, and DP, mention should be made of a popular criticism of pacifism. This criticism, as stated, for instance, by Elizabeth Anscombe, is that pacifism amounts to an indefensible "clean hands policy." The pacifist refuses to take the brutal measures necessary for the defense of the nation, for the selfish sake of maintaining his own inner moral purity. It's argued that the pacifist is thus a kind of free-rider on the rest of us, gathering all the benefits of citizenship while not sharing all its burdens. The pacifist is not willing to fight, yet gains security from those who are. This picture, as drawn, is of a rather unlikeable, holier-than-thou, selfish free-rider. A related inference drawn is that the pacifist might actually be an internal threat to the security of the state, because his non-participation and self-absorbed objection diminishes the state's resources and moral resolve in fighting.[25]

This "clean hands" argument can be overstated, especially in countries with challenging national security situations, like Israel, wherein mandatory military service is required of all citizens, and the precariousness of the overall political situation may motivate accusations of the selfish pursuit of "clean hands."[26] But it's important to note that, to the extent to which *any* moral stance will commend some actions deemed morally worthy, and condemn others as being reprehensible, the "clean hands" criticism is so malleable as to apply to any substantive moral and political doctrine. *Every* moral and political theory stipulates that one ought to do what it deems good or just, and to avoid what it deems bad or unjust. *The very point of morality itself, we might say, is to help keep one's hands "clean."* So, this popular criticism of pacifism is not especially appealing. Besides, the very idea of a selfish pacifist simply does not ring true in more general contexts: many pacifists have, historically, paid a very high personal price for their pacifism during wartime (through severe ridicule, job loss, ostracism, even jail time), and their pacifism seems less rooted in regard for inner moral purity than it is in sincere regard for constructing a less violent and more humane world order.[27]

3.2 Describing and Evaluating TP: War at Odds with Virtue and Human Flourishing

Teleology is a basic perspective on the nature of morality itself. Literally meaning "the study of purpose," teleology often now gets referred to as "virtue ethics." Virtue ethicists, such as Aristotle, believe that human beings must live their lives trying to develop their innate capabilities to the fullest extent. This, for them, is the purpose (or *telos*) of life. Of course, we have many capabilities to develop: intellectual, physical, social, and so on. What does it mean to develop one's *moral* capacity to the fullest? It's to pursue ethical excellence, which is displayed by the virtues. What are the virtues? They are freely chosen character traits which we praise in others. We praise them because: (1) they are difficult to develop; (2) they are corrective of natural deficiencies (e.g., industriousness is corrective of our tendency to be lazy); and (3) they are beneficial both to self and society. The virtues, Aristotle insisted, often fall within a mean between extremes of behavior. His favorite example was usually courage, which he defined as the virtuous mean between the "deficient" extreme of cowardice and the "excess" extreme of being rashly risk-prone and foolhardy. (Many of his examples of courage, incidentally, deal with courage related to warfare and battle.) The reason for the insistence on the mean is that, according to Aristotle, we observe that, when a creature's behavior inclines toward the middle ground across a range of options, the creature is thereby strengthened and tends to flourish to the maximum extent possible.[28]

There are many virtues, and a moral person is one who develops them and displays them consistently over time. In this sense, the virtues are like muscles in that they need habitual conditioning to be strong and "toned." The ancient Greeks listed four "cardinal virtues" – wisdom, courage, moderation, and justice – and Christian teaching is well known for its recommendation of faith, hope, charity, and love. Other prominent virtues include honesty, helpfulness, forgiveness, pleasantness, consistency, tolerance, thoughtfulness, and so on. For the human vices, simply negate the above. These are destructive of self and society; give in to easy or even wicked instincts; and represent a triumph of the base, mediocre, nasty, and corrupt.[29]

The essence of teleological pacifism (TP) is this: when we think of warfare and war-fighting, we see that none of this is praiseworthy activity. Violence, killing, and bloodshed seem clearly at odds with the kind of ideal life – a fully realized and excellent human life – which is the focus of virtue ethics. War is not part of any sane person's idea of a flourishing existence. Warfare causes terrible pain, loss, and suffering, and the inflicting of violence brutalizes the victim and arguably causes a corruption of character – a hardening and insensitivity – even to the inflictor. A TP pacifist would also suggest that, although some aspects of courage might (admittedly) be called upon in war, just as common is the experience of post-traumatic stress disorder (PTSD), which reduces the formerly strong soldier to a broken shell. Moreover, which is truly more courageous: fighting, or refusing to fight in spite of the danger? There are a number of sharp questions here, connected to core virtues: is war truly a wise choice? A moderate and humane one? One expressive of hope and charity, or rather of hatred and malice? Isn't war, as a destroyer, the very opposite of creativity and life? How can war be consistent with love? Indeed, doesn't all this show that *peace itself is a virtue* – part of the human ideal – and that, although it might be very difficult, it's in pursuit of peace that we nevertheless must always orient our thoughts and actions?

A compelling set of considerations. How have some argued against them, or at least raised critical questions? One issue commonly expressed, whether by realism or JWT/LOAC, is whether the commitments of the pacifist are utopian – i.e., excessively unrealistic. War might indeed be a rough business which, ultimately, calls forth more vice than virtue. But it might also simply be needed to defeat a violent, aggressive, rights-violating regime: like Nazi Germany or Imperial Japan. It might be said that a world where such aggressors are allowed to triumph, and then inflict brutality, is not part of any sane person's idea of the best life, either. And there may be something to be said for: the virtue of defending one's people (or fellow citizens) from conquest and enslavement; the courage it takes to confront a ferocious aggressor; the self-discipline and group organization it takes to fight smartly and decently; and of the strength and ingenuity it takes to formulate and execute a successful war plan. More generally, JWT/LOAC would assert that *justice is also a virtue*, and a major one at that. Much of the dispute between pacifism and JWT/LOAC has to do with different perspectives on *what to do when the world presents us with situations where the virtues of peace and justice are in conflict* and cannot both be realized. Where just war theorists differ from pacifists is that they suspect that there can be cases where justice does not include peace or, more precisely, where peace may include or enable injustice. Michael Walzer, for example, argues that, by failing to resist aggression with effective means, pacifists actually end up *rewarding* aggression and *failing* to protect people – fellow citizens – who need it.[30]

Pacifists reply to these arguments by contending that we don't need to resort to war in order to protect people and to punish rights-violating aggression effectively. In the event, for example, of an armed invasion by an aggressor state, an organized and committed campaign of non-violent civil disobedience – perhaps combined with international diplomatic and economic sanctions – would be just as effective as war in expelling the aggressor, with much less destruction of lives and property. William James famously labeled such a committed campaign "the moral equivalent of war."[31] The conviction is that no invader could possibly maintain its grip on the conquered nation in light of such systematic isolation, non-cooperation, and non-violent resistance. How could it work the factories, harvest the fields, run the transportation network, man the stores and banks, when everyone would be striking, refusing to comply, or quietly sabotaging orders? How could it maintain the will to keep the country in the face of crippling economic sanctions and diplomatic censure from the international community? And so on. Peter Ackerman and Jack DuVall have detailed many actual, historical cases of non-violent resistance around the world; Robert Holmes and Gene Sharp, amongst others, have developed further some of the abstract non-violent tactics which pacifists might rely on.[32] Consider the following list of resistance tactics offered by Sharp, and note how they, at the very least, call into question the tiny "tool-box" of a mere three basic foreign policy options (of diplomacy, economic incentives, and sanctions) prior to the use of armed force:

> general strike, sit-down strike, industry strike, go-slow and work to rule ... economic boycotts, consumers' boycott, traders boycott, rent refusal, international economic embargo and social boycott ... boycott of government employment, boycott of elections, revenue refusal, civil disobedience and mutiny ... sit-ins, reverse strikes, non-violent

obstruction, non-violent invasion and parallel government.[33]

Though one cannot exactly disprove this pacifist proposition – since it's mainly a counter-factual thesis – John Rawls suggests (and no doubt realists of all stripes would agree heartily) that this may be "an unworldly view" to hold. For the effectiveness of these imagined campaigns of civil disobedience *would depend upon the standards and scruples of the invading aggressor.* But what if the aggressor is utterly brutal, ruthless? What if, faced with civil disobedience, the invader "cleanses" the area of the native population, and then imports its own people from back home? It's hard to strike, or sabotage orders, if one is dead. And what if, faced with economic sanctions and diplomatic censure from a neighboring country, the invader decides to invade *it*, too? We have some indication from history – particularly that of Nazi Germany – that such pitiless tactics are effective at breaking the will of even very principled people to resist. The defense of our lives and rights may well, against such heartless invaders, require the use of political violence. Indeed, under such conditions, Walzer says, adherence to pacifism might even amount to a "disguised form of surrender." He thinks that, in such instances, if one truly believes in values like "resisting aggression effectively" and "protecting oneself and fellow citizens from aggression, or genocide," then one should be willing to fight for them – because, in such cases, *fighting holds the only realistic prospect of actually defending these values.* Unwillingness to fight here translates into non-support for these values. But how can you not want to protect people, and to resist aggression, and attempts at ethnic cleansing and genocide, effectively?[34] Ward Churchill and Mike Ryan actually say that pacifism is "a pathology of the privileged": i.e., something advocated by those who are secure, whereas those under grievous threat of enslavement or death simply cannot afford such pampered belief.[35]

Pacifists respond to this accusation of "unworldliness" by citing what they believe are real-world examples of successful non-violent resistance to rights-violating aggression (and, further, effective resistance *by the weaker against the stronger*). Examples most often mentioned include: (1) Mahatma Gandhi's campaign to drive the British Imperial regime out of India in the late 1940s, leading to the independence of modern India; and (2) Martin Luther King Jr.'s civil rights crusade in the 1960s on behalf of African Americans.[36] Walzer replies curtly that there is no evidence that non-violent resistance has ever, *of itself*, succeeded. This may be rash on his part, though it's clear, for example, that Britain's own exhaustion – financial and otherwise – after World War II had much to do with the evaporation of its Empire and the eventual independence of India. Walzer's main counter-argument, against these so-called counter-examples, is that *they actually illustrate his point*: that effective non-violent resistance depends upon the scruples of those it's aimed against. It was only because the British and the Americans had some scruples and standards, and were in the end moved morally by the determined idealism of the non-violent protesters, that they acquiesced to their demands. But realism and JWT/LOAC argue that *aggressors will not always be so moved*. It might seem unthinkable to us, but a tyrant like Hitler, for example, might interpret non-violent resistance as disgusting weakness, deserving contemptuous crushing. "Nonviolent defense," Walzer suggests, "is no defense at all against tyrants or conquerors ready to adopt such measures."[37]

To those pacifists who retort that even Hitler was faced with his own non-violent resistance – namely in Norway and Denmark after he conquered those lands in 1940 – the point gets made that the problems that the Norwegians and Danes put in his way because of their strikes, sabotage, and protest cannot really be considered successful acts of pacifist resistance to aggression. They happened, after all, *after* Hitler had already conquered those lands – so how were they helpful in resisting the aggressive takeover of their countries? Second, perhaps the reason why the Nazis didn't crush these Scandinavian protest movements were: (1) they had already conquered Scandinavia, and didn't consider these protests a serious threat to their control; and (2) they now had bigger fish to fry, like England, Russia, and America. Third, and speaking of the major powers, *they* were the ones who beat Hitler – *with force* – resulting amongst other things in the liberation of Scandinavia. It was *not* home-grown pacifist protest which got the Nazis out

of Oslo and Copenhagen; it was the decisive military defeat of
Nazi Germany by the remaining Allies. This is not to deny the
fact that Scandinavian resistance *did* cause problems for the Nazis,
that it boosted local spirits, and that the co-ordination required
was impressive, the acts often brave and ingenious. It is, rather, to
put them into their proper perspective as smart and bracing tools
of resistance but not, ultimately, as tools successful in rolling back
Nazi aggression.[38]

Walzer contends that James' idea of an effective "war without
weapons," much less a world without war, is (for now) a "mes-
sianic dream." Realists of all stripes would, of course, join with
JWT/LOAC in agreeing. For the foreseeable future, and in the real
world we all inhabit, it's better to be prepared to resort to armed
force in the face of armed aggression or attempts at genocide. This
doesn't have to imply a willingness to use any and all means at
one's disposal, though we've seen that classical realists may recommend such. Structural realists, and certainly JWT/LOAC, instead
urge restraint in fighting: restraint aimed at limiting war to the
achievement of such compelling objectives as the defeat of regimes
like those of Nazi Germany and Imperial Japan. "The restraint of
war," Walzer opines, "is the beginning of peace."[39]

Here's another way philosophers have of making this point:
there's an important difference between "ideal theory" and "non-
ideal theory." Ideal theory is what's true, or best, under ideal
conditions: i.e., a world with endless resources, nothing but wise
people, and with everyone wanting and trying to be morally
good. Under such perfect conditions, it may be irresistible to
conclude that the only permissible policy on these issues would
be pacifism. But the world is not ideal. In particular, there are
bad people and nasty regimes who indulge in violence and seek
out rights-violating domination over others, or sometimes even
their systematic slaughter. It's this very sad and sub-optimal fact
about our world which gives rise and plausibility to the use of
protective armed force, whether seen as smart self-serving strategy
(by the realists) or else as a full-blooded moral entitlement, and
perhaps even an obligation, by JWT/LOAC. Both realism (by
definition) and JWT *are pieces of non-ideal theory*: sets of ideas and

values designed to guide us in the here and now, in the real world, until deeper and deeply better transformations can be found and generated. Pacifists, of course, would view such "accommodation to reality" as a form of giving up, and would either remind us of their ultimate (often religious) ideals or, at least, of William Morris's interesting argument, at last chapter's end, regarding how the world is ever supposed to improve unless we hold on to our ideals and refuse to surrender them to a rough-and-tumble world. It's a foundational difference in world-view.[40]

3.3 Describing and Evaluating CP: War's Costs Always Exceed Its "Benefits"

Consequentialism is another basic account of the essence of ethics. As the name indicates, the focus of consequentialism is on the concrete results of one's actions. This tradition is skeptical of the value of focusing on personal character traits, as teleology does, or on abstract universal rules (as we'll see deontology does). The key, ethically, is whether the world ends up better as a result of one's actions. The proof is in the pudding: the right thing to do, in every instance, is to perform the action that is going to have the best contribution to the world's overall welfare. So, consequentialism involves: (a) a serious attempt to predict the costs and benefits of one's options for action; and then (b) a mandate to act in accord with the one promising the highest "payoff" (in terms of pleasure, happiness, or welfare) to the world at large.[41] The CP element of contemporary pacifism, accordingly, is the notion that *the costs of war always outweigh the benefits from undertaking war.* One doesn't need to be a military historian to get the point, namely, that war is incredibly destructive, brutal, often appallingly inhumane and shockingly cruel. And the benefits, if any, of wars are often seriously unclear and are soon overtaken, in any case, by the flow of historical events. As the famous song asks: "War, what is it good for?" The CP pacifist agrees with the next line, "Absolutely nothing!"[42]

At that level of generality, such a claim seems difficult to deny.

One potential controversy, which realism and JWT/LOAC might mention, is that we should consider not only the explicit costs of war action (i.e., both military and civilian casualties, the costs of deployment, the destruction of property), but also the implicit costs of war inaction: not resorting to war may be tantamount to rewarding aggression – armed rights violation – in international relations. The lack of armed resistance and forceful punishment might allow an aggressor state to keep the fruits of its campaign, thereby augmenting the incentives in favor of future aggression. To what extent can we have a well-functioning and stable – much less a just – international system in which aggression between nations is so rewarded? Call this the "macro-cost of war inaction": the rewarding of interstate aggression, which leads to the long-term weakening of the interstate system of peaceful dispute resolution.[43] The thought experiment becomes grimmer for pacifism when we think not just of classical cross-border wars but of potential armed humanitarian interventions (AHI). How the costs of inaction here – as tragically witnessed in Rwanda in 1994 – aren't simply sovereignty and land but literally hundreds of thousands of lives. We'll detail the case in the next chapter, but here put forth the query: *would non-violent non-co-operation have saved the Tutsis from genocide?*

We must also talk of the "micro-costs" of not resorting to war to defend one's own people from an aggressive invader. Such a pacifist strategy seems, at the least, to run enormous risks with the safety and well-being of one's citizens, not to mention their right to be self-governing and not subject to a conquering regime. In light of this, we should ask: does pacifism make sense at the level of collective agents, like states, *especially* if those collective agents are charged with the responsibility of protecting and serving their citizen members? Is there too great a reliance, by the pacifist, on the intuitive appeal of the inter-personal case (of not killing another individual), as opposed to the analogous, yet different, political or international case? In other words, pacifism at the level of the individual – the conscientious objector, for instance – seems much more plausible and principled than advocating that an entire state be geared along pacifist lines, *especially given the sub-optimal situation of the international arena* (as the realist would remind us).

Again, the risks to one's people would be huge, and so it's deeply unclear whether a prudent, or moral, government ought to adhere to such a view. Former US president Barack Obama declared as much in his 2009 Nobel Peace Prize speech, acknowledging on the one hand the great achievements of Martin Luther King Jr. and Gandhi, yet at the same time pronouncing: "[A]s a head of state sworn to protect and defend my nation … I … cannot stand idle in the face of threats to the American people."[44]

The combination of both the macro- and the micro-costs of *failing* to resort to war (in some well-defined situations) may be sufficient at least to cast doubt on whether war-fighting can *only* result in greater negative, than positive, consequences. It's simple to cluck one's tongue and shake one's head at the destruction of warfare − "War is bad!" − and quite another to think through the costs of pacifism and what *they* might involve. Being against war, after all, can be like being in favor of motherhood − a "no-brainer" that no one in her right mind would disagree with. But it's different for those charged with the responsibility to protect their people, especially when threatened by a terrible aggressor who relishes warfare and seeks out conquest over weaker others. As one of my (pacifist) students once reluctantly commented: "The *nice* answer might not actually be the *best* answer." No realist could have said it better![45]

We might also, in our consideration of CP, consider the historical record. It does indeed seem true that many − perhaps even most − historical wars can be objected to, very forcefully, by the CP aspect of contemporary pacifism. World War I appears a fitting example of the futility, waste, and sheer human tragedy of many wars our ancestors fought.[46] But not all wars seem to fall neatly under this objection. World War II, for instance, is more debatable. Many thoughtful people, including participants who actually made the sacrifices, have argued − appealing to both prudential and moral costs − that defeating ultra-aggressive regimes like Nazi Germany, Fascist Italy, and Imperial Japan was worth the costs of the war-fighting, as enormous as those admittedly ended up being. Can we, they ask, imagine and endorse what our world would currently look like had the Nazis been allowed to conquer Europe and rule it, had Mussolini spread his "New Roman Empire" beyond Ethiopia, and/or had Imperial Japan been allowed to subdue most of East Asia? George Orwell's searing image of a soldier's boot "stomping on a human face – forever" comes to mind. World War II didn't create a wonderful world – the world of our dreams – but it did prevent a truly terrible world from coming into being.[47]

A third issue to raise with regard to CP focuses on the relationship between consequentialism and the denial of killing, especially on the level endemic to warfare. Pacifism places great, perhaps overriding, value on respecting human life, notably through its usual injunction against killing. But this core pacifist value seems to rest uneasily with the appeal to consequentialism in CP. For there is nothing to a *consequentialist* approach to the ethics of war and peace which would *always* outlaw killing. There is here no firm, absolute principle that one must never kill another person, or that nations ought not to launch military campaigns that kill thousands of enemy soldiers. With consequentialism, *it's always a matter of considering the latest costs, benefits, and circumstances.* Consequentialism was actually first designed, quite explicitly, to be flexible in a way in which teleology and deontology aren't. (Note the difference between consequentialism and realism: while both appeal to cost–benefit utilities, consequentialism remains an *ethical* doctrine focused on the world's overall improvement, whereas realism is a doctrine *skeptical of ethics* and thus featuring strong commitments only to one's own selfish benefit, as a tribe, group, or nation-state.[48])

Since it's always a matter of choosing the best option amongst feasible alternatives, consequentialism clearly leaves conceptual space open to the claim that under *these* conditions, at *this* time and place, and given *these* possible alternatives, killing and/or war seem(s) permissible. There can be particular counter-examples, some offered above, wherein the calculations come out the other way. After all, what if killing 2,000 people (say, some members of an invading army) seems necessary to save the lives of 9,000 people (say, people of one's nation who would die from the rights-violative activities of the unchecked invader, as it sought to consolidate its rule)? *It's at least possible that a quick and decisive resort to war could be employed effectively to prevent even greater suffering, killing, and devastation in the future.* Indeed, military and political historians often engage in this thinking, and sometimes claim things like: "Had the Allies confronted Hitler after Austria, it wouldn't have taken so long, later, to defeat Germany. Appeasement made the war longer and more destructive." To put it plainly, it seems bizarre for the consequentialist pacifist, whose principles exhibit a profound abhorrence for killing people, to be willing in such a scenario to allow an even greater number of people to be killed by acquiescing to the violence of others less scrupulous. Two related points, of potential critical objection, are being made here: (1) the general point that the CP element of pacifism does not, of itself, seem to ground the categorical, absolute rejection of killing and war which is the very essence of pacifism; and (2) the particular point that CP seems open to counter-examples (like World War II) which question whether consequentialism would even reject killing and war at all in certain conditions. Consequentialism might actually *recommend* warfare, if the circumstances were dark enough and the other options limited enough.⁴⁹

Pacifists would no doubt reply that all these concerns under-describe the full and manifold costs of war and the enormous drain on our society by such massive "investments" in armed conflict. They might refer back to the mention of the military–industrial complex in chapter 1. And, as to "the costs of war inaction," the debate would revert back to that regarding the potential of effective non-violent resistance, considered above in section 3.2.

3.4 Describing and Evaluating DP: War Violates Basic Moral Duties

Deontology is yet another major world-view regarding morality. Deontology has, as its core intuition, the notion that the concept of duty is at the foundation of ethics. Ideas like duty, obligation, and responsibility are uniquely moral ones – indeed, the most uniquely and clearly moral ones. There might be confusion between morality and mere personal prudence when we talk, like teleologists, about beneficial character traits or when we speak, like consequentialists, about costs and benefits. But there's no mistaking that we're talking about morality when we consider duty and obligation. So, it's thought that these ideas, and not the others, must be the essence of ethics. Doing one's duty is what's central in morality and justice. And by "duty," deontologists usually mean that one's behavior is permitted, or demanded, by a first principle or general rule regarding morality, such as "Thou shall not lie," or "Honor your father and mother." How does deontology get used by pacifists?

The exact nature of the DP element varies from thinker to thinker, but the core notion is that *the very activity of war-fighting violates a foremost duty of morality*. Thus, undertaking such activity can never be justified by appealing to the aims or consequences of the war action in question. War, as a means to an end, is thought to be intrinsically unjust. The supposed "justice" of the goal sought, through war, does not redeem the injustice of the means used to pursue it. There must be consistency between means and ends. War ought never to be resorted to: there's always some vastly superior option with regard to international dispute resolution, such as diplomacy, sanctions, or organized campaigns of non-violence.

Now: *why* does war-fighting violate *which* foremost duty of morality? Perhaps the best pacifist consideration of these matters has been offered by Robert Holmes. Holmes deals with this consideration, at first cut, by stating that the foremost duty of morality violated by war-fighting is *the duty not to kill other human beings*.[50] But the obvious objection to this claim is that such a duty – though crucially important under normal conditions of life – does not seem to override all other considerations under special, very threatening circumstances. Consider the most obvious example: A brutally attacks B, thereby posing a severe threat to B's life. Provided that A attacks without justification, many people would respond that B may retaliate against A in self-defense, with lethal force if needed. Consider another example: are we not to kill a dangerous terrorist who is credibly threatening the lives of many innocent civilians? Say, one of the 9/11 hijackers while he was in the act? It seems defensible to assert that we may, so that we protect the lives of the innocent.[51]

What about one of the highest-profile moral duties, The Golden Rule itself: i.e., to do unto others as you would have done unto you? *Since no one wants to be targeted with killing force, one shouldn't do so to anyone else.* Good (though such comes from religious and not secular pacifism), yet a possible counter might be to say that it crucially depends upon which starting point we're focusing on. If the Nazis (or terrorists) have already invaded or attacked, and are setting out to kill me, then they've broken The Golden Rule themselves – and have been the first to do so. Rights of self-defense may thus re-appear. And I would want, if I could, to save others from killing violence: if someone else (like my government) can do the same for me, then why shouldn't they?[52]

DP pacifists are not, at this point, out of options. Holmes, for example, proceeds to suggest that the foremost duty of justice violated by war is not the duty *not* to kill aggressors or terrorists but, rather, *the duty not to kill innocent, non-aggressive human beings.* To be innocent here means to have done nothing which would justify being harmed or killed; in particular, it means not constituting a serious threat to the lives and rights of other people. It's this sense of innocence that JWT/LOAC invokes when it claims (as we'll see, in fuller detail, in chapter 5), that civilians should not be directly attacked during wartime. Even if civilians support a war effort politically, or even in terms of their personal attitudes toward the war, they clearly do not pose direct and serious armed threat to others. Only armed forces, and the political–industrial–technological complexes which guide them, constitute serious threats against which threatened communities may respond in kind. Civilian populations, JWT/LOAC surmises, should be off-limits as targets. Holmes, as well as Cheyney Ryan, contend that this JWT/LOAC rule of non-combatant immunity *cannot ever actually be satisfied.* For all possible wars in this world – given the nature of military technology and tactics, the heat of battle, complex fighting environments, and the limits of human knowledge and self-discipline – involve the killing of innocents, thus defined. We know this to be true from history and have no good reason for expecting otherwise in the future: *all wars kill some people who didn't deserve to die*. But the killing of innocents, Holmes says, is always unjust. There's no more basic moral principle than that! So, no war can ever be fought justly, regardless of the nature of the goal sought after, such as national defense from an aggressor's attack. The very activities constitutive of war violate bedrock ethical principles and cannot be redeemed by the supposed justice of the ends they are aimed at.[53]

This objection is perhaps pacifism at its morally strongest. To adjudicate the issue, though, we need to develop the JWT/LOAC perspective more fully. Why? Because JWT/LOAC is going to assert that, provided that the rules of *jus in bello* (of permissible warfighting) are otherwise met, then some kinds of civilian casualties *may* be permitted or excusable. In particular, if one has made every reasonable effort not to harm civilians, yet some such harm still happens, plus if one is fighting an otherwise justified war to begin with, then *some* unintended, indirect, "collateral" civilian casualties *might* be permissible *in light of* the overall importance of the war effort (say, to beat the Nazis, or to save the Tutsis). Obviously, to consider such issues fully, we need to consider what counts as a justified war, and a justified way of fighting such, from the perspective of JWT/LOAC. So, this is what we turn to next, leaving pacifism as nicely developed from an initial perspective, but awaiting further comment in the following chapters at the appropriate moment. Another further issue, to be tabled and returned to, is this: we can examine pacifism, as we have in this chapter, from the perspective of individual ethics and either the morality or rationality of collective choice. But another way to frame it is through the longer-term, institutional lens of making the international system more peaceful overall, both for its own sake and to save individual states, non-state actors, and individuals from the sharp moral dilemmas we've encountered here. There is, for example, something called "the perpetual peace tradition" which has offered a number of institutional proposals for achieving just that. It's the very embodiment of cosmopolitanism, as it purports to offer ways in which the entire world can be structured so as to guarantee the end of warfare, which in many ways would be the ultimate pacifist dream.[54] But it seems best, after having put it here on the table, to come back to such ambitious, "end-state" proposals only at the very end of this book, once everything is fully in view.

3.5 Conclusion

Pacifism occupies the ethical, idealistic, cosmopolitan extreme of opinion on these issues, in stark and nearly complete opposition to realism. The realist would thus accuse the pacifist of being a naïve "do-gooder," a utopian with a big heart but also a bad case of wishful thinking – so severe, in fact, that it may lead to very dangerous risks to the security and well-being of one's political community. JWT/LOAC would, in a sense, go even farther and say that the risks of being inactive and passive, or engaging only in non-violent resistance, may actually violate the most basic duty of a government, namely, to do what it can to protect the lives and rights of those it has responsibility for, and especially from the most ferocious aggressors who have intentions either to slaughter or to enslave. Pacifists deploy a variety of strategies in response. In realism, pacifism sees nothing but the "same-old, same-old": a bleak and brutal cycle of never-ending violence and constant jockeying for position and power which history demands we escape from, and this can only be done by refusing to play along, and insisting on a totally different, and more elevated and idealistic approach, such as non-violent (or, at least, non-warring) resistance. Against JWT/LOAC, pacifism deploys a number of counter-arguments, designed to call into question who has a better understanding of morality and justice. Here in this chapter we canvassed three such large arguments: that war-fighting is at odds with human virtue and flourishing, that it's never worth the costs it imposes, and that it violates some of the most foundational duties that human beings have. Rejoinders of a kind were developed, especially for the first two accusations, but the outcome regarding the third demands now a full development of JWT/LOAC.

4

Just War Theory and International Law: Start-of-War

Just war theory (JWT) and international law (especially the so-called "laws of armed conflict" (LOAC)) together form the third major, systematic perspective on armed conflict. As stated in the introduction, they are *not* the exact same thing; *yet* they overlap substantially and share *both* the same method – of applying abstract universal rules of conduct to warfare – *and* the same core value convictions (amongst others) that war sometimes might be permissible, and there's a difference between permissible and impermissible means of fighting it.

JWT is much older than the LOAC, by hundreds of years, and fittingly has had a long, complex development, drawing its values and understandings from such diverse sources as religious writings, traditional practices, ethical values, political debates, and direct military experience (especially that surrounding controversial wars, which spark deeper thinking about: "Is this a justified war?" "Was that the right weapon or tactic to use at that time?" "Who is to blame for all this?") The LOAC, by contrast, were deliberately constructed in the modern era by national governments and, when they began to do so, it was natural to turn to JWT for some guidance: no point in re-inventing the wheel when there was a pre-existing body of thought. But we'll see that, even though they are closely entwined, JWT and the LOAC aren't completely the same thing, and can come apart in important ways.

The main aim in this chapter isn't to develop the long history of either tradition. There are other, excellent sources to which the reader is referred in the note.[1] Here we're concerned with the abstract principles: their meaning and rationale, and their application to concrete cases of political violence. We'll refer to "the history of ideas" only when helpful. And we'll move from the classic accounts, based on assumptions of state-vs-state symmetrical war, to more contemporary applications involving asymmetrical war, wherein non-state actors and complicated, multi-party, multi-phase armed conflicts increasingly texture our world.

4.1 Standard, Classical *Jus ad Bellum*

It's easiest to begin with the rules of the LOAC. Why? Because as experts say, international law *merely stipulates, or declares,* certain rules of conduct. It does *not* explain or justify them: it simply declares the consensus agreement of what national state governments are willing to abide by and acknowledge. It's precisely JWT, as a moral theory, which strives to offer compelling justifications for the principles; moreover, JWT usually endorses rules above-and-beyond those of the LOAC.[2] This is quite typical, if you think about it, as morality usually holds people to higher standards than does the law (e.g., infidelity within a romantic relationship is widely criticized ethically, but is usually not punishable by law). There can be further issues of the morally ideal demanding things about which, strictly, it may be very difficult to gather good evidence (which would be needed for any kind of legal standard of action). So, the law shies away from endorsing those principles. We'll see this, for example, when we consider "right intention" in wartime. Keeping these things about JWT/LOAC interaction in mind, let's quickly stipulate the LOAC rules, and then consider the JWT additions to such rules and, above all, the JWT explanations and justifications.

The LOAC stipulate that countries may resort to armed force *only if all* of the following four criteria of "*jus ad bellum*" ("the justice-of-war") are met:

- just cause
- proportionality
- last resort
- public declaration of war, by the proper authority for doing so.

A head of state, or whomever holds "the war power" in the country in question (see below), must ensure that these criteria have been satisfied before embarking on war; otherwise, they may wind up charged with war crimes after the armed conflict. Standard JWT accounts of *jus ad bellum* commonly add two further rules: right intention and probability of success.[3]

4.1.1 Just Cause

This is the most important rule, by far, and sets the tone for everything else that follows in wartime. As important as it is, though, we should stress immediately that all the other *jus ad bellum* rules *must still be satisfied as well*: too many people forget the other criteria once the complex debate over just cause has been thrashed out.

Most experts agree that, when it comes to a just cause for going to war, international law (especially as codified in the United Nations Charter, articles 39–53, particularly 51–3) recognizes three general principles:

1. All countries have the inherent, "natural" right to go to war in self-defense when facing aggression. Aggression is defined in international law as the first use of armed force against another country.
2. All countries have, in addition, the inherent, "natural" right of other-defense – otherwise known as collective security – to go to war as an act of aid to *any* country victimized by aggression.
3. *Any other use of force,* such as a pre-emptive strike or an armed humanitarian intervention (both considered below), is *not, in*

the eyes of international law, an inherent, "natural" right of states. Any country wishing to engage in these more controversial forms of force must get the *prior* approval of the United Nations Security Council (UNSC). Failing to receive prior authorization renders any such use of force illegal, itself an act of aggression.[4]

So, if Country A commits an armed attack against Country B (without prior approval of the UNSC), then B (and any other country C, D ... or Z) is entitled to go to war against A as an act of *defense from*, *resistance to*, and *punishment of*, aggression. Aggression is seen as a wrong so severe that war is a fitting response, based on the fact that aggression violates the most basic rights of groups and individuals to life and security, and to freedom and well-being – that is, to go about their lives peacefully on a territory where their people reside and not be otherwise killed or enslaved to a foreign power. Classic examples of international aggression include the following:

- Imperial Germany's invasion of Belgium in 1914, sparking World War I
- Nazi Germany's invasion of Poland in 1939, causing World War II
- Japan's invasion of China in 1937, and its attack on the US at Pearl Harbor in 1941, triggering the Pacific War during World War II
- the Soviet Union's invasion of Afghanistan in 1979
- Iraq's invasion of Kuwait in 1990, touching off the Persian Gulf War.

There are thousands of historical examples of international aggression, and it generally tends to involve a bigger, bellicose state invading a smaller, less powerful one in order to grow its own power, prestige, and resources.[5]

Let's now consider – by way of moving beyond the legal declarations into the moral explanations of such – the justice of an armed response by a victim (V) of aggression against its aggressor

(A). This is important, not only to provide rationale for the law but for the clash with pacifism. Usually, five major reasons are offered:

Reasonableness. It would be *unreasonable* to deny V permission to take effective measures to protect its people from serious harm, or lethal attack, at the hands of A. It's not reasonable, given the kinds of creatures we are, and the kind of world in which we live, to expect a state charged with the responsibility to protect its citizens instead to capitulate utterly – to just roll over – in the face of aggression. Indeed, one of the core functions of a government is precisely to protect its people from foreign invasion and accompanying threats of slaughter or slavery. As the ancient Greek philosopher Aristotle – probably the coiner of the phrase "just war" – memorably declared, "it is morally obvious" that a community has the right to defend itself from attack.[6]

Fairness (and Deterrence). The *fairness* argument stipulates that it would be not only unreasonable but unfair to deny V the permission to resist A with force, should A unjustly attack. Why? Because, in the absence of such measures, V will suffer substantial loss of life and liberty while A will actually *gain* whatever object it had in mind in attacking V. It seems clear that aggression ought not to be unfairly rewarded in this way, both in the particular case and for the sake of not eliciting more such behavior in the future – and that's the connection to *deterrence*. Strong resistance deters, or probably prevents, future aggression.

Responsibility. Not only would denying V permission to resist forcibly the aggression of A be unreasonable and unfair, it would also ignore the fundamental issue of *who bears responsibility* for the choice situation. It's the aggressor A who's responsible for placing V in a situation where V must choose between its rights, its people, and its interests, and those of the aggressor A. (And this point ties nicely into the reasonableness claim, since it's only reasonable to expect V, in such a situation, to choose its own.) The aggressor A, if it doesn't wish to experience defensive force by V, *can always cease and desist* from its own aggression. Stop warring, if you don't want to be warred upon! So, if A does *not* cease and desist – if it keeps up its aggression – it's hardly in a position to cry foul should

V decide to do the reasonable and fair thing and defend itself, and its citizens, with the needed force.[7]

Implicit entitlement. Finally, any victim V has the *implicit entitlement* to use whatever measures are necessary to realize the objects of its rights and those of its citizens. It's an axiom of modern political theory that human beings have rights (or justified claims, or elemental entitlements): in particular, to life, liberty, the means for sustaining such, and further core norms of non-discrimination and recognition as a human being and rights-holder. People form countries and governments to better realize and secure these individuals rights, especially against domestic criminals and foreign aggressors.[8] So, countries themselves have rights – state rights – and these are recognized in international law. The two major state rights are those of *political sovereignty* and *territorial integrity*: i.e., the right to govern oneself without foreign interference (assuming one reciprocates) and the right to do so in a secure physical space, wherein one is further seen as entitled to the natural resources within that space, to meet the needs of one's people. International aggression is, more precisely, *the use of physical force in violation of state rights to sovereignty and integrity*.[9] The notion is this: when faced with aggression, countries cannot count on any effective world government to step in and deal with the law-breaker, so to speak, the way that individuals can in well-governed societies (thus saving those individuals from having to fight for themselves and their rights). There *is* the UN, true, but it's *not* an effective world government. And there *are* other countries who might be one's allies and they *may* come to one's aid if one is attacked – but *maybe not*, depending on their own interests and capabilities. When push comes to shove, in the face of aggression in our sub-optimal world, countries often have to rely on themselves for defense and, in the last resort in any case against the most ferocious and determined aggressors (like Nazis or mass-casualty terrorists), armed self-defense may well be required. One must have the implied right to use those means (armed defense) truly necessary to secure one's more basic rights (to life and liberty at the individual level, or security and sovereignty at the national level). It's moral logic: if I have an original right to Q, I must also have the further right to whatever is genuinely needed to have and secure Q.[10]

We know that many pacifists will strongly object and sweepingly declare that "two wrongs don't make a right," in referring to the situation between aggressor (A) and victim (V). But JWT and the LOAC emphatically do *not* view it the same way, and thus there's a substantial divergence of opinion here which roots the different theories: JWT/LOAC *agree* that the first action, aggression by A, is wrong; but they *disagree* that a defensive use of force in response by V (or any third-party helper, TPH) is a second wrong. That second thing, rather, is *a morally justified and legally permissible defensive use of force*, for the five reasons given above of reasonableness, fairness, deterrence, responsibility, and implicit entitlement. The potent nature of these moral reasons further leads just war theorists to view realism as *under-describing* the situation here: yes, they'd generally agree with the realists that it's probably in one's rational self-interest to defend oneself from attack. But *it's even more than that*: it's morally justified as well. Hence, "just war theory."[11] (We'll delve further into the important issue of non-defensive force below, after the remaining *jus ad bellum* criteria have been explained.)

4.1.2 Proportionality

In every kind of law, there is supposed to be a proportion, or balance, between problem and solution (or between violation and response), which is to say that the LOAC commands that the problem in question be serious enough that war is a proper, fitting reply. Since war is so costly, bloody, and unpredictable, it follows that only a very few problems indeed are truly so bad that war will ever be a proportionate response. What problem, if any, is truly so severe that war is a genuinely proportionate response? International law's answer, for JWT reasons stated above, is *aggression*: when confronted with an aggressive invader – like Nazi Germany, Imperial Japan, or the Soviet Union – which is intent on conquering and essentially enslaving other nations, it's deemed reasonable to stand up to such a dark threat to life and liberty and to resist it and beat it back, with force if need be. Just as dangerous criminals must be resisted and prevented from getting away with their crimes – lest chaos be invited – countries are entitled to stand up to aggressors, and to defeat them.[12]

4.1.3 Last Resort

But one is first supposed to have "exhausted," or made all best attempts, at first having resolved the issue in question – including aggression – with the other tools of foreign policy as we described them in chapter 2. Obviously, if one can solve a problem through diplomatic negotiation, or through economic "incentivizing" (either positive or negative), then that is much preferable – including to oneself – than running the sizable risks and costs of armed conflict. (And we recall, from the last chapter, how pacifists call into question the smallness of that tiny list of but four items/ options in "the foreign policy tool-box.")

It's important to note, as Michael Walzer has explained, that we are not talking about "literal last-ness" here. While there's always *something else* which could be tried (such as one more round of talking), the meaning of this rule is relative to the full context of the situation: the gravity of the threat, the nature and actions of the aggressor, and the preferences and capabilities of the victim and any of its allies. As hum-drum as it sounds, the norm of "what's reasonable," given the full description of the situation, is what JWT and the LOAC commend as being key. *Is this use of force going to be reasonable?* We know this is, for them, going to be relative to those foundational moral reasons stated above in the just cause rule, whereas for the pacifist, this standard will never be satisfied. The perhaps more pointed or helpful re-phrasing by JWT/LOAC would then be: will *not* resorting to force *lead to* unreasonable, unacceptable, unjust consequences ... such as failure to protect those who need protection, loss of vital resources, massacre or genocide, loss of sovereignty, or enslavement to a foreign power with terrible intentions?[13]

A useful illustration of this principle concerns the run-up to the Persian Gulf War of 1991. We've mentioned already that, in

August 1990, Saddam Hussein's Iraq invaded its tiny neighbor Kuwait. International allies, led by the US and Britain, first tried to talk to Saddam. They tried to negotiate with him, then they tried threatening him, to no avail. They then slapped sweeping sanctions on him, trying to prevent him from profiting from his new-found, ill-got, oil gains (and got many of his neighbors to do likewise). Still nothing. The allies then made a big show of slowly creating a huge build-up of military force along Saudi Arabia's borders with Iraq and Kuwait ("Operation Desert Shield"). Saddam wasn't intimidated. As a result, the international community felt the decision to go to war, to physically push Saddam's military out of Kuwait and back within Iraq's own borders was, indeed, a last resort. This they did ("Operation Desert Storm") within 2 months, in early 1991.[14]

4.1.4 Public Declaration of War by a Proper Authority

War is supposed to be declared out in the open, officially and honestly, by the individual or government department/ministry with the authority for doing so. The origins of this rule go all the way back to ancient Roman times, and are connected with such figures as the Roman Senator Cicero (106–43 BCE), who was insistent that Rome's many wars of imperial conquest needed to follow proper rules of governmental procedure: in particular, to try to prevent such ultra-ambitious Roman generals as Sulla and Caesar from gaining too much power, and essentially using public military forces for their own private advancement and greed.[15]

In every country nowadays, some branch of government has *the war power*: the authority to order the use of force and warfare. Generally speaking, in many democracies, the legislature has the war power, whereas in most non-democratic societies, it's the executive – the president or dictator. In Canada, for example, the war power rests with Parliament; in the United States, the war power likewise seems to rest, constitutionally, with the legislature (i.e., Congress). But the American president, as commander-in-chief of the US armed forces, has enormous factual power to order the American military into action. As a result, many experts argue that the war power in the US is actually split – in classic American "checks-and-balances" style – between the legislative and executive branches of government. This arrangement produced an epic struggle between the branches during both the Korean War (1950–3) and especially the Vietnam War (1954–74), when Congress felt that successive presidents were running a *de facto* war without actually declaring it and getting *de jure* (i.e., legitimate) authority for doing so from Congress.[16]

More recently, following the start of the Global War on Terror (GWOT) in late 2001, Congress has extended to the various US presidents since then rather sweeping *advance permission* to deploy force as they best see fit to counter terrorism, in particular with an eye to empowering rapid response. So, there have been fewer controversies in this regard of division-of-power in our own era: however, as experts note, perhaps the real power which Congress exerts here is over the military budget, and it can always slash resources there if it feels the president has made a severe, lingering error. Two vital topics where norms of publicity and authority remain relevant in our time concern the use of drones within the GWOT, and cyber conflict. The temptation to use new forms of force in ways which may be secret, or evade public scrutiny, is historically a very seductive temptation for those with the war power; and the main function of this rule is to try to block such abuses. We'll return to those two specific subjects in subsequent chapters.[17]

4.1.5 *Two Just War Theory Add-Ons*

On top of these four legally codified principles of *jus ad bellum*,[18] JWT stipulates two further conditions: right intention and probability of success.

The rule of right intention was invented by Augustine (354–430), who's often (incorrectly) reported as the inventor of all JWT. He's important, and he did create *this* rule, but he's hardly the founder of the whole theory: we've already witnessed contributions from both Aristotle and Cicero who, after all, lived hundreds of years before Augustine. Augustine was both a public official and a religious authority, and during his lifetime the Roman Empire was starting to fall apart in Western Europe, especially from repeated attacks by Germanic tribes. This concerned Augustine greatly, as these tribes were not Christian but pagan, and he worried that, should Rome fall apart, Christianity might cease to exist: who would carry on the legacy (e.g., by teaching the doctrine to future generations)? As a public official, then, he believed that perhaps it *was* a good idea to use the might of the Roman military to defend the Empire and its citizens from unbelieving barbarians. But Augustine was also Christian; and, as we saw in the last chapter, there's not much evidence in the New Testament for concluding that Jesus permits violence and killing. Augustine wrestled anxiously with how to square this circle, and came up with the following: *it's a Christian ruler's duty to love and protect his people*, emphatically those who are both innocent and cannot defend themselves. In the real world – the sub-optimal "City of Man" – this duty trumps the general Christian disposition against violence. But a ruler ordering war must do so *only* out of such love, and *only* with the greatest reluctance: his/her personal intentions or subjective motives must thus be pure and unsullied by such typical human weaknesses as greed for resources, or hatred for the enemy. Love, protection, defense from aggression: these are the only permissible motives behind a decision to go to war.[19]

Why not include this JWT rule of right intention in the LOAC? There is thought, in the first instance, to be a problem with evidence: how to know for sure the motives and intentions of such enormous, complex, multi-part entities as national governments and their military establishments? This is especially biting when there are competing factions within such, as was the case within the US federal government in connection with the Vietnam War and the second Iraq War (2003), the controversial "anticipatory attack" which resulted in regime change. Second – and I suppose this comes from realists – assumption is made that "purity of motive" *might* be something we require of individual human beings but *cannot* reasonably or plausibly expect from countries caught in a complex, rough-and-tumble world wherein mixed motives are almost inevitable.

On the one hand, these reasons make sense, in terms of the exclusion of right intention from the LOAC. On the other, ordinary people *do* commonly use this norm when discussing the ethics of war: it was, for example, very common for America's 2003 Iraq attack to be criticized on the grounds that President George W. Bush's "real motive" was "to get the oil." And, within domestic law, intention not only appears as a factor in some crimes (e.g., often the difference between kinds of manslaughter or murder precisely has to do with different intentions on the part of the accused), it's assumed that one's intentions *can be empirically inferred from* one's behavior. Yet it remains a simple fact that right intention, while a part of JWT, is not an element within the LOAC.[20]

The same holds for the rule of probability of success. The moral notion here is that it's wrong to go to war if you know, in advance, that it's likely that you'll lose. To do otherwise is to engage in death and destruction that will make no difference to the outcome: something which seems both irrational and insensitive to human suffering. This sounds compelling; but the reasons for this rule's exclusion from the LOAC are three-fold. First, sometimes the outcomes, both of particular tactics and entire wars, are not so predictable. States and non-state actors can make mistakes, even severe ones; new weapons and relevant technology can be invented anytime, and shift momentum; and unusually gifted soldiers, officers, or political leaders can make a big difference. (Winston Churchill, for example, defied the views of many in the UK political establishment at the time by insisting that war was the only way to stop the Nazis, and that such a war could be won.) Second, we can't forget that the LOAC are contractual deals between national governments seeking benefit: probability of success is thought to be biased toward big and powerful countries (as, odds are, they'll likely win their wars) and so, small and medium-sized countries refuse to ratify any treaty requiring such a rule. Don't smaller countries, they ask, also have the sovereign right to engage in armed defense against aggression?[21] Finally, recall from chapter 1 that the very meaning and metric of "success" can vary widely between actors – notably between, say, the guerilla war tactics favored by insurgents and terrorists versus the conventional tactics preferred by official state militaries. Sometimes, "success" can be in the eye of the beholder: a military loss may turn into a political victory (as a community may long celebrate its failed resistance against a much larger power as an act of tragic but "identifying" defiance, as depicted, say, in the film *300* regarding brave but doomed Spartan attempts to hold off an enormous invasion of ancient Greece by the Persian Empire).[22] Or, a short-term military victory may turn out to be so costly that it leads to long-term military and political failure. This is referred to as a "Pyrrhic victory" after the experience of Pyrrhus of Epirus, who held off the ancient Roman armies twice, but at such great cost in casualties to his side – compared to the Romans, who could field many more replacements much more quickly – that he famously said that "one more victory like this, and we shall be uttered ruined." And he was.[23] More recently, we might reflect on the American experience in Afghanistan (2001–) and Iraq (2003–): on the one hand, militarily successful in the strict short-term sense of overthrowing both regimes, which was in the first instance the objective in both cases; on the other, a very difficult, long, struggling "post-war" period, after which it's not at all obvious whether, on the whole, the Iraqi or Afghani people, or even the American people, would consider those two armed conflicts to have been "worth it" or "successful."

4.1.6 Big Picture Synthesis

We should note, using helpful concepts from the last chapter, that the ethical structure of the *jus ad bellum* within the JWT is a blend of both consequentialist and anti-consequentialist elements. An anti-consequentialist approach to ethics submits that the rightness or wrongness of an action can be determined even before we know its results: for example, if it violates somebody's basic rights, or it violates a foundational moral duty. We know such an action is wrong – as it breaks a very compelling first principle – even before we see that it may also, for example, produce pain and suffering. Anti-consequentialism thus puts stress on proper procedure (as opposed to results), and the importance of there being a fair set of rules that are binding on everyone. We see such a perspective contained in the three JWT *jus ad bellum* rules of: just cause, right intention, and public declaration by a proper authority. Go to war only to resist aggression (either against oneself or others); do so with the proper motive (precisely of trying to stop and defeat the aggression, so that you protect people); and be sure to be open, public, and accountable about it. And you only get to set all this in motion if you are the proper, legitimate authority for doing so.

Conversely, a consequentialist approach to ethics stresses the over-riding importance of the consequences, or results, of one's actions, and views *those* as being indicative of whether an action is right or wrong. We see concern for generating good consequences (or, at least, avoiding or minimizing bad ones) in the three JWT *jus ad bellum* rules of: proportionality, last resort, and probability of success. Make sure war is in fact proportionate to, or truly called for by, the problem; make sure to try all reasonable non-violent or alternative measures to solve the problem first (as they will probably be much less risky and costly than war); and be very confident that, in taking the dramatic step toward political violence, one has some chance of success in achieving one's objective, else one's manifold efforts will be futile – and much pain and sacrifice will produce nothing of value.

It's thus reasonable to believe that JWT, and the associated LOAC, aspire to be comprehensive, "common-sensical," middle-of-the-road, accounts of warfare which blend two of the most fundamental approaches to morality and ethics: we need to evaluate armed conflict both from the *before-hand* perspective of rights, procedure, and fair first principles, as well as from *after-the-fact* considerations of costs, casualties, and sufferings, and odds of success or failure.

4.2 Non-standard, Contemporary Amendments to *Jus ad Bellum*

Four pressing subjects fall under this heading, as they don't fall neatly into the classical account yet, increasingly, compose much of contemporary armed conflict. They are:

- The Global War on Terror (GWOT)
- Anticipatory attack or pre-emptive strike
- Armed humanitarian intervention (AHI)
- Justice during civil war

4.2.1 The GWOT's Beginning in Afghanistan

Terrorism, and responding to it, are important sub-texts throughout this book. For our purposes here, in connection with the *jus ad bellum*, we mention two things: (1) how non-state actors, like terrorist groups, *can commit aggression as readily* as can state governments and their militaries; and (2) something called "state-sponsorship of terrorism." We know the GWOT begins as a response to the 9/11 attacks by the terror group al-Qaeda on America in 2001, killing more than 3,000 civilians in New York, Washington, and Pennsylvania.[24] These actions by al-Qaeda fulfill the definition of aggression as offered above: a deliberate, armed and killing attack across borders, in violation of a country's right to political sovereignty and territorial integrity and of the human rights of its people to life and liberty. An armed response, on the part of the USA, was thus warranted against al-Qaeda in terms of just cause. So, why did America invade Afghanistan?

The answer is that the national government of Afghanistan at the time was the Taliban. They shared al-Qaeda's violent Islamic extremism and, in fact, used their control of Afghanistan to enforce a very strict, idiosyncratic interpretation of religious law within that country. So, the Taliban were natural ideological allies with al-Qaeda. More than that, though: they had provided al-Qaeda a "safe harbor" within the borders of Afghanistan, allowing the terrorist group to live and operate there, with training camps in particular. This constitutes *state sponsorship of terrorism*: when an official state government provides safe harbor, weapons, training, intelligence, resources, and deliberate general enablement and non-prosecution of terrorist groups, the idea is that such a government *is itself complicit in, and guilty of,* any act of terrorism and/or aggression which they may commit. The analogy here is to when someone knowingly and willingly "aids and abets" a criminal within domestic society: they are also guilty of wrong-doing. The moral relationship is thought to be transitive: if A commits aggression against B, and C has deliberately enabled A to do so, then C has also committed aggression against B. B may thus, with just cause, reply with defensive force against either A or C, or both: and other countries may join in with B as acts of other-defense/ collective security. This is precisely what happened when the US invaded Afghanistan in October 2001, leading to the collapse of the Taliban government about seven weeks later. The US was joined by 28 allied countries in the action, Operation Enduring Freedom.[25]

Such analysis, though, *only* applies to the just cause rule within the *jus ad bellum*: we recall that all the other criteria must also be satisfied. In the America/Afghanistan case, public declaration by proper authority was certainly met; and presumably so, too, was right intention in terms of avenging the 9/11 attacks and sending strong signals in favor of deterring future attacks, with no questionable ulterior motives like resource grabs (as Afghanistan has very little). But opinions differ regarding the remaining criteria:

- In terms of proportionality, was invading and overthrowing another country's national government actually proportionate to the terrorist crimes of killing 3,000 civilians (albeit in gruesome, high-profile, fashion)?
- In terms of last resort, some note the rather quick timing of a mere month before invading, and wonder whether other options for dealing with al-Qaeda had truly been exhausted (e.g., no sanctions were placed on Afghanistan). Others may reply, though, that the US had in fact asked the Taliban to hand over al-Qaeda personnel and shut down the terrorist training camps, and were ignored. Such people might also question what "last resort" is actually supposed to mean when you've been murderously attacked in such dramatic, large-scale, aggressive fashion: 9/11 was this generation's Pearl Harbor (Bin Laden apparently said that 9/11 was designed to be such). And so, having been forcefully attacked, there was little else reasonable to do but reply with a forceful counter-attack – lest one seem

completely incapable or unwilling to protect one's people. The government of the United States cannot allow it to be "open season" for terrorists to hunt down American civilians.[26]

• And we've already raised issues of probability of success: not so much the short-term objective (very quickly secured indeed, of forcing the Taliban from power); but, rather, the longer-term calculation of what was to be done *after* the Taliban fell and the US *still occupied* Afghanistan, with responsibilities therefore to help re-build. Could they, at the start, have reasonably predicted a successful post-war exit, or not?

4.2.2 *Anticipatory Attack, Especially on Iraq*

The GWOT underwent a significant – perhaps sideways, or even backwards – development with America's invasion of Iraq less than two years later in May 2003. The invasion is important not only for that reason but also for problematizing the issue of whether any non-defensive use of political violence might ever be justified. Can an anticipatory attack, or pre-emptive strike, ever be legitimate? We've seen that the LOAC offers a purely procedural answer: yes, but only if there has been a prior author-izing vote allowing for such by the UNSC. Was such UNSC authorization offered to America for the 2003 Iraq invasion? No: after Russia and China made clear that they would veto any such petition for authorization, the US declined to bring the resolution to a vote – but proceeded to invade anyway, for reasons detailed below. As a result of this move, of proceeding without legal blessing, almost no international lawyers – not even American ones – will say that the 2003 Iraq war was permitted by the LOAC. Many references abound, in this regard, to "an illegal war."[27]

There's a difference of opinion within JWT regarding anticipa-tory attack. Some are what we might call "strict defense purists," who insist that morally you must wait until you are actually attacked with aggression, and *only then* are you entitled to respond either in self- or other defense. To do anything else would be, in the memorable words of Spanish Renaissance theologian Vitoria, "to punish a man for a sin he has yet to commit." Strict defense purists will sometimes further say that anticipatory attack is, by definition, at odds with the other *jus ad bellum* rule of last resort: it's difficult indeed to claim that you've exhausted all other options when you're the one who strikes first.[28]

Opposing the defense purists in this regard are just war revisionists. We shall see throughout chapters 4–6 that there is an internal "team tension," or "sibling rivalry," between these two groups: more traditional, conventional just war theorists with quite conservative and literal readings of just war and international law rules; and the revisionists, who think that *the principles behind* the rules are more important, and accordingly have a broader, more liberal reading – with some quite interesting and even surprising implications. One of them is allowance, in principle, for anticipatory attack. How do they justify this?

Jeff McMahan, a talented and influential revisionist, says that it's crucial to note that defense has actually never been entirely passive, reactive, and backward-looking. Even when an aggressor A has already attacked a victim V, a major objective in V's resort to defensive force in reply has always been *to prevent yet more harm* at the hands of A and *to make A stop its aggression*. And so, it has (*pace* Vitoria) never truly been "traditional" for defense merely to imply "the second party to use force." The essence of defense, rather, has always been about *protection*: the protection of *lives* as well as the protection of such vital *values* as freedom and security. It can be consistent with this general ideal of protection that you may have to use a blend of first-strike and secondary-reply measures, depending on the case.[29]

Consider, by way of analogy, the police authorization to use force in domestic society. National and local police forces are not strictly required to wait for the criminal suspect to perform his crime or attack. If he's behaving in a very threatening way, is armed, shows clear signs of wanting (and being about) to use such, then *they are entitled to use force first* in order to take him down before he does even greater damage. And – this is the crucial thing – when done properly, this is all done in the name of, and for the sake of, defense: protecting the public from the danger posed by the criminal. Where there is, as famously formulated, "a clear and present danger" of significant, rights-violating harm, then the first use of protective, yet still genuinely defensive, force may be called for.[30]

This is to say that the JWT revisionists must have a slightly different concept of aggression, at least compared to the LOAC. They would assert that aggression is *not* adequately, or comprehensively, defined as it is in international law, namely, as "the first use of armed force across an international border." This, they would say: (a) imbues borders with far too much moral value; and (b) rules out, by conceptual fiat, the existence of aggression *within* a country. Their preferred, alternative definition: aggression is the deliberate use of armed force in violation of the most basic rights of persons and countries, in particular to freedom and the physical security needed to exercise that freedom. Importantly, this more expansive definition means that *aggression can happen internally*. State governments can commit "internal aggression" against their own people as well as "external aggression" against foreigners. The advantage of making such a move is that it makes sense of, and provides rationale for, armed humanitarian intervention (AHI) as well as for speaking about justice during (at least some) civil wars. It's a broader definition, and yes, for some traditionalists/purists/literalists that may be concerning. But it's also much more robust and allows us to speak intelligently about all these non-traditional cases, which we know are increasingly the face and future of warfare itself. Contrast that with strict, traditional readings of international law, which suffer the huge drawback of being unresponsive to, and uninformative about, these non-classical cases of armed conflict.[31]

More in a moment about AHIs and civil wars. Let's finish our examination both of the general principle of anticipatory attack, as well as the particular recent case of America's pre-emptive strike on Iraq in 2003. Walzer, the dean of living just war theorists (and usually of a more traditional kind, but on this specific subject quite revisionist), proposes three criteria which ought to guide any justified use of anticipatory attack by one country, X, against another, Y:

1. Y has shown "a manifest intent to injure" X (shown either through recent, explicit threats, or a long-standing bitter mutual conflict). Y is, objectively, "a determined enemy" of X.
2. Y is making active military preparations which make the intent "a positive danger" (e.g., by amassing armed forces along a mutual border).
3. X's "waiting, or doing anything other than fighting, greatly magnifies the risk of suffering attack" by Y.

Walzer even goes so far as to say that, if all the above criteria are met and the national government of X does *not* engage in a pre-emptive strike, then it's actually *derelict in its duty* to protect its people from aggression and harm.[32] As stated above, in connection with the deep reasons behind a defensive resort to force, one of the most basic things we look for, from any kind of government, is reliable protection from domestic criminals and foreign attackers: if a government cannot provide these to us, on a secure and regular basis, then it's not at all clear why we should obey such a government and view the social contract as beneficial and binding. How is it "my" government if it cannot, or will not, protect me?

It's an illuminating illustration to consider whether America's 2003 Iraq attack meets Walzer's criteria for a justified pre-emptive strike, as something like this *was* the main argument offered by the Bush Jr. administration at the time. The way the administration saw it was this. In terms of condition one, where X represents America, there were two relevant Ys: the government of Iraq, as led by Saddam Hussein, and the terrorist group al-Qaeda. We can agree with Bush that both of these groups were, indeed, "determined enemies" of the US at the time of the 2003 attack: al-Qaeda had just attacked America during 9/11, a year and a half prior; and Iraq was still under sweeping sanctions from the end of the Persian Gulf War in 1991, and deeply resented that.[33]

In terms of conditions 2 and 3, the Bush administration's argument was this: there was a conspiracy plot by Saddam to give some of his weapons of mass destruction (WMD) to al-Qaeda, for them to use on American soil, resulting in another 9/11-scale terrorist attack with mass casualties. We know that al-Qaeda would have eagerly done so, keen on "keeping up their momentum," or even topping the brutal body count they achieved during 9/11. As for the immediacy of any such conspiracy threat, the argument at the time wasn't firmly made, rather more an assertion as to "we can't just sit here and let that happen ... as it could happen at any time." The notion was well put in the 2002 *National Security Strategy of the United States*: "The greater the threat, the greater the risk of inaction – and the more compelling case for taking anticipatory action to defend ourselves, even if uncertainty remains as to the time and place of the enemy's attack. To forestall or prevent such hostile acts by our adversaries, the United States will, if necessary, act pre-emptively."[34]

The formal *structure* of the argument seems sound, but it's the factual *content* which is profoundly in question. Not so much the first condition, as granted, but rather the second two. As most Middle East experts said at the time, and since: there's no doubt that, separately, both Iraq and al-Qaeda were indeed "determined enemies" of America. However, this did *not* make them allies even in an abstract sense, much less in the very strong and concrete sense of hatching a collective plot to use WMDs to attack America in another 9/11. Why not? Because Saddam was a secular dictator. Throughout his reign in Iraq, Saddam vigorously opposed and attacked radical Islamic extremists like the people in al-Qaeda. He knew that, if they could, they would overthrow him, and install a radical Islamic government in Iraq: he feared this very much, as, for example, it had happened right next door to him, in Iran, in 1979. Thus, the notion that he would hand over WMDs to al-Qaeda is widely agreed to have been nonsense, as they would have just as readily used them *against him* as against the United States. It would have been an insanely risky move, one that not even his hatred of America could render plausible. The subsequent, official Congressional investigation into the 9/11 attacks went through all the evidence and determined that there was *no evidence whatsoever of any kind of conspiracy agreement*, or any alliance at all, at any time, between Saddam's government and al-Qaeda. Since that's the crucial step in the argument – the "active preparations" between the co-conspirator Ys for an "imminent attack" on X – on the whole the justification for the 2003 Iraq attack must fail, at least according to Walzer's criteria (and, as we saw previously, in terms of international law as well, regarding the non-authorization by the UNSC). This is probably why there was only one other country, the UK, who agreed to go along with the US in 2003, as opposed to the 28 who joined together in 2001 in Afghanistan. (It's clear that anticipatory attack is always going to be much more controversial than defensive force.[35])

Here's the tie-in to the claim regarding the GWOT going sideways or backwards during the Iraq invasion. It's quite possible that what led US decision-makers to see conspiracy where there was none *was precisely the intense GWOT fear of being subject to another 9/11*, and an over-riding concern to avoid "9/12" (as they now say), almost at all costs. Bush Jr. may even (as someone with an MBA) have thought of it in risk-management terms: "Even if I don't have clear evidence of any conspiracy, I just cannot take the chance that there *might* be one. I'm better off 'taking the war over there,' overthrowing Saddam, and gaining control over those WMDs so that they can't be given to anyone." And then that decision being correct, as far as it went, but then leading to severe embarrassment once no WMDs were found, and then severe further problems, to be discussed in chapter 6, of American occupation and post-war reconstruction in Iraq, which cannot plausibly be said to have advanced the broader GWOT objectives at all. If anything, the chaos in post-war Iraq encouraged domestic terrorism within the country between rival groups there, and led to gaps in control and policing which, by default, allowed terror groups safe harbor, including, for example, some of those who eventually coalesced into ISIS/ISIL, and took over chunks of northern Iraq from 2014 to 2017.[36]

4.2.3 AHI and Civil War: Rwanda, Libya, and Syria

While not logically necessary, it may be appropriate – in today's context[37] – to consider these issues together, as many real-world cases in recent times have involved complex mixtures of both civil wars between local groups fighting for control of their country, as well as external, international intervention within such civil wars, in the name of rescuing one or more groups from utter ruin and slaughter at the hands of a rival group. Consider, for example, the Rwandan genocide.

During its colonial days under the rule of Belgium, Rwanda was dominated by a minority group, the Tutsis. The majority group, the Hutus, resented being discriminated against and, when independence from Belgium came in 1959, the Hutus turned the tables and took control. Relations between the groups were always tense in subsequent years, at times breaking out into armed conflict between the Hutu-dominated Rwandan Army and the Tutsi-dominated rebel force, the Rwandan Patriotic Front (RPF). In 1993, after several years of fighting, a peace accord and power-sharing arrangement were achieved. Before they could be implemented, Hutu extremists formed private militias to carry out an audacious and murderous plan: not only to destroy the peace accords but to eliminate the Tutsis and even any moderate Hutus who supported peace with the Tutsis. They truly had genocide on their minds: attempting to kill an entire people. The Hutu extremists seized power in a *coup d'état* in early 1994, killing the moderate Hutu leadership. While the UN did have a peacekeeping force already in Rwanda (there to help enforce the 1993 accords), as soon as some of its (Belgian) members were killed in the coup's early days, the UNSC ordered a complete withdrawal – in spite of the pleas of the peacekeeping unit's own commanding officer, the Canadian Romeo Dallaire. He knew what was going to happen and presented the UN with detailed evidence of a genocidal plot. But he was ignored, over-ruled, and all Western troops left Rwanda, taking their own nationals with them. The Hutu extremists had free reign to execute their horrible plan and, by April, had killed about 800,000 people, both Tutsi and Hutu. Most were butchered brutally with machetes or shot at point-blank range with small guns. *Fully one-third of all Tutsis on earth were murdered.* France, alone amongst Western countries, re-intervened late in June and helped re-establish some sanity (though some say France allowed the perpetrators to leave the country as well). Then – then! – the UNSC voted to return "peacekeeping" forces back to Rwanda.[38]

Many agree now that Rwanda was a near-perfect moral case for when AHI *should* have happened, but tragically *didn't* for various reasons. It should have happened because, in terms of *jus ad bellum*, there was: (a) just cause (saving a comparatively helpless group from attempted genocide); (b) proportionality (of armed intervention as being proper and balanced in light of trying to stop violent genocide); (c) a high probability of success (in that the "*genocidaires*" were only lightly armed in military terms, and so putting a sophisticated armed force between them and their targeted victims would've stopped the fighting – as, indeed, France eventually proved); and (d) last resort (in that once large-scale violent massacre is under way, what other option is there, save to stop the attackers in the quickest, most efficient way possible, literally with physical force). Had the decision to engage in AHI been made, of course it could have been publicly declared by a proper authority, and right intention presumably would've been satisfied, as Rwanda's lack of strategic location or huge quantity of valuable resources would've meant that, had an AHI happened, the sincere motive really could only have been to rescue the people there from murderous massacre.[39]

Realists point out that this last item actually shows why AHI didn't, in fact, occur in Rwanda in time to prevent the mass slaughter: it not being in the clear self-interests of any of the major powers at the time to get involved in Rwanda, beyond the moral one to save the Tutsis from genocide. Realists further opine, as we know from chapter 2, that countries require substantial strategic reasons of self-interest to involve themselves in something as costly and risky as armed conflict: morality and justice alone, they say, are nowhere near enough. Never have been, never will be: and sometimes, that can lead to dreadful results for (usually weak and vulnerable) groups and nations, like the Tutsis. Whether that's true or not as a descriptive analysis – I suppose it's "true enough," in terms of there being many historical cases like that – we should comment further on the morality or justice of AHI.[40]

The notion, from a JWT perspective, hinges on the above idea of "internal aggression": if one group is deploying killing force in violation of another group's basic human rights to life, liberty, and security of the person (and, emphatically, if that first group is also in control of the state government and its military assets, whereas the second group is defenseless, or severely out-gunned), then the first group *is precisely an internal aggressor*. And we know from our analysis of aggression that: (a) the targeted/victimized group may permissibly fight back in self-defense; and (b) any other group/country/countries willing and able to take on the burdens of other-defense is likewise entitled to do so. This would be the revisionist JWT understanding of the moral legitimacy of using armed force in order to intervene in a foreign country. Rwanda is also frequently cited by JWT as a real-world case against pacifism, in the sense that non-violent resistance by the Tutsis against the Hutu extremists would've failed to prevent genocide, whereas armed force would've stopped it – as France's actions eventually proved.[41]

In 2000, the government of Canada – feeling its peace-keepers in Rwanda, as led by Dallaire, were betrayed by the UNSC – convened the International Commission on Intervention and State Sovereignty (ICISS) to try to better guide the UNSC regarding when it ought to authorize AHIs. The "responsibility to protect" (or "R2P") doctrine, outlined in the Commission's final report, asserts:

1. All states have the responsibility to protect their own people from such "mass atrocity crimes" (MACs) as: genocide, war crimes, crimes against humanity, and ethnic cleansing.
2. If a state *fails* in this responsibility, then other states have *the duty to step in*: (a) in the first instance, to aid and enable the state's capacity, if lack of resources is the issue; or (b) to intervene with armed force, if the issue is (rather) the state itself turning murderously against its own people.[42]

Crucially, R2P is not deemed a violation of state sovereignty, as it's explicitly stated that state governments *only have rights* of political sovereignty and territorial integrity *when they fulfill their own duties* to protect their people from MACs. It's the most basic political exchange, in social contract terms: I'll obey your laws as a government, but only if (at the very least) you agree to protect me from MACs. We really cannot require anyone, in terms of basic rationality, much less morality, to obey a government engaging in, or allowing, a mass atrocity crime aimed against them. Clearly, such a view is both an expression, and an endorsement, of a revisionist JWT understanding of defense, aggression, and proper authority.[43]

Though R2P is *only* considered a new norm (or value, or expectation), and *not* full-blown international law, some experts have argued that it may be on its way to becoming the latter. After all, the UNSC itself endorsed R2P, in 2011 when it authorized an AHI in Libya.[44]

Fighting broke out in Libya in late 2010 as part of the so-called "Arab Spring" uprising. (This refers generally to sweeping anti-authoritarian and pro-human rights protests and changes which have been sweeping the Middle East and North Africa (MENA) region since late 2010, with different results in different countries.[45]) The fighting in Libya coalesced into two groups: those supportive of existing dictator Muammar Gaddafi; and those devoted to his overthrow. It was civil war. When Gaddafi's army turned violently against not only the rebel groups but unarmed civilians, and even whole towns deemed to be his "enemies," NATO – initially led by France and Italy, which have historical ties to Libya – sought authorization from the UNSC to intervene with armed force. In March 2011, the UNSC gave its approval – citing the R2P doctrine as rationale. From March to October 2011, NATO forces provided aid to the rebels, and performed many direct strikes – especially air strikes – on their own against Gaddafi's forces. America, Britain, and Canada were robust participants in this action, which ended when Gaddafi was killed in October. The NATO mandate ended in November 2011, and was arguably successful as an AHI in the sense that large-scale, deliberate massacre of civilians by the Libyan government was stopped. More broadly, though, the new government replacing Gaddafi wasn't able to stay stably in power, and Libya remains in a complex condition of post-conflict transition, with rival groups (including iterations of both al-Qaeda and ISIS) still jockeying for control.[46]

The case of intervention in Libya stands in stark contrast to the non-intervention in Syria, which also saw fighting erupt between pro- and anti-government forces in the wake of the Arab Spring. And Syria's president, Bashar al-Assad, responded the same way as Gaddafi had done: a mega-violent crackdown on all his perceived enemies, and refusal to change or reform, much less share power or resign. Assad's forces have not only engaged in deliberate civilian massacre, they've even deployed prohibited WMD, such as chlorine "barrel bombs" and even sarin nerve gas. Syria's very bloody, ongoing civil war – estimated casualties of 500,000 and growing, alongside approximately 5 million refugees fleeing the fighting – has been raging, with no AHI undertaken, or even contemplated, by NATO or the UNSC more broadly. Since the two cases of Libya and Syria seem so similar, when thus described in terms of their origins, what marks the difference: AHI in one, nothing (thus far) in the other? The reason seems to be that, in spite of their deep *moral* similarities, there are profound *political and strategic* differences between the two cases, hence tying into a contrast between JWT and realism. The Libyan regime under Gaddafi, for example, had few if any allies in the region, whereas the Syrians have tight ties with Iran and (UNSC veto-holder) Russia. Libya is also a wide-open, mainly desert, country, whereas Syria is tiny and densely populated. So, an air-based intervention in the former was easily do-able (high probability of success), whereas any intervention in the latter would probably involve many civilian casualties and involve complex fighting in dense urban environments: one of the least-favored fighting options (low probability of success). Finally, AHI in Libya allowed the West to show support for regime change in the Arab Spring and, having done that, the case for doing it again – in a far riskier environment – was much lessened. Others, more realist, have also noted how France and Italy were deeply interested in the Libyan oil supply (whereas Syria has none), and America and Britain had anti-terrorist scores to settle with Gaddafi going back to the 1980s. Whereas intervention in Libya led to government change – though of a non-ideal kind – non-intervention in Syria seems to mean that Assad's government will probably be able to weather the severe storm of civil war and hold on to power. We'll see. And, beyond the politics, there's the principle, and deliberate non-intervention in a case where R2P may have been demanded only calls into deeper question the true nature and binding import of this norm in the future.[47]

4.2.4 *Political Legitimacy: The Right to Govern and Use Force*

This brings us back to a fitting final subject in this chapter: political legitimacy. Perhaps the ultimate philosophical question in connection with the start of war (civil, international, or otherwise) is precisely: *who has the right to govern this group of people, on this territory, and why?* For, if A *does* have the right to govern a given people and territory, then others (either internally or externally) do *not* have the right to resist its rule (or, certainly not with large-scale killing, in any event). Whereas, if A does *not* have the right to govern a given people and territory, then it is the one who has *no right to complain and resist should others* (again, either internally or externally) defy its governance, perhaps with armed force if that is truly needed as a last resort to get rid of its ghastly grip on power.[48]

This bedrock issue of political legitimacy gets handled quite differently from either a traditional or revisionist perspective, and we note how it's one of the most profound issues of political theory in general. The traditional, international law account – heavily influenced by history, and realism's impact on such – has tended to assert that *political legitimacy has to do with whichever group can actually, effectively exercise power and control* within the territory in question. This group will, almost by definition, control that country's state government, and so then things like civil war, attempts at revolution and secession, as well as local massacre, are precisely going to be seen as "internal matters" about which other peoples should mind their own business – unless and until, as a matter of fact, a new group can come to claim effective control of the state. To the upside, this traditional view is: quite clear; can be empirically determined (i.e.: who does, in fact, control that country?); and is even comparatively simple and would liberate us from worrying about the justice of all the messy cases of armed conflict we've just mentioned. The only metric which matters is substantial, de-stabilizing violence that crosses a pre-existing international border. The downside, though, is that this traditional view is incomplete, seems seriously out of date, and demands silence in connection with issues about which many informed people don't want to be silent, such as whether Assad should be in control of Syria, or whether terrorists should be allowed safe harbor in some far-away land, or whether we should just stand idly by while genocide is attempted against the Tutsis, to the tune of approximately 800,000 murdered.

From the revisionist JWT perspective, the traditional view incorrectly collapses issues of legitimacy and authority into ones of (mere) power, and is thus barely distinguishable in the end from realism. Whereas power is the raw, actual ability to get what you want, or to perform some action, authority is defined as *the right to use such power*, the entitlement to perform that thing in question. Authority and legitimacy inescapably involve judgments about morality and justice, and so *cannot* be reduced to mere observations about power and control. Hence, revisionist JWTs often speak of requirements of "minimal justice" in connection with the political legitimacy of a state and its concomitant right not only to govern but to deploy force (if needed) for the sake of protecting its people. Though many differ on exact details, there is a rather large consensus that most revisionist JWTs share in this regard, and it leans in the following direction.[49] A state government is legitimate, and thus has the right both to govern and not to be attacked with force, provided:

1. *It is non-aggressive, in both the internal and external senses of that word* (i.e., it doesn't commit or allow MACs on its own people; and it doesn't attack foreign countries in violation of the rules of *jus ad bellum*).
2. *It is seen as being the legitimate government of that society, both in the eyes of its own people, as well as the international community.* The international community extends this recognition by allowing the society membership in various international associations and events, for example, the UN or the Olympics. Individual countries also do things like send ambassadors to, and open

embassies in, that society as well as enter into contractual agreements with it, involving trade, for example. The society isn't shunned on a widespread basis nor is it the subject of pervasive official skepticism about its fitness to organize life in its community. Domestically, the most obvious way for a people to show that it recognizes its own government as legitimate would be through majority endorsement in periodic, public, free, and fair elections. It's not the only way – simply the most obvious and verifiable, the "gold standard" (as it were) of domestic legitimacy.

3. *It makes every reasonable effort to satisfy the human rights of all its citizens.* Human rights are elemental entitlements to, or justified claims on, those objects we both vitally need to live a minimally good life and whose provision or enablement would impose only reasonable duties or burdens on the rest of society. I argue that there are five foundational human rights objects in this sense: life or personal security from violence; individual freedom; material subsistence; equality, at least in the form of non-discrimination; and recognition as a person and rights-bearer.[50]

Such a more robust conception of political legitimacy is required to lift JWT and international law above and beyond realism, as well as to enable the judgment of all the non-traditional cases of armed conflict analyzed above in this section. (For example, the key *jus ad bellum* question in connection with civil war, in general, will therefore be: which side, if any, has minimal justice on its side?) But such a conception comes at the cost of asserting values which may be contested. It is impossible to avoid this, however, given the moral standard it wishes to uphold. It tries to mitigate this problem by keeping the values "minimal" and "thin," and as widely-agreed-to as possible: (a) in the form of basic human rights for everyone; (b) in the conviction and observation that aggression is wrong and seriously destabilizing; and (c) in the common-sense view that the government of a people should be accepted by them, and be seen as their legitimate representative by most of the international community.

4.3 Conclusion

This chapter opened our development and analysis of the JWT/
LOAC perspective on war and political theory. We developed the
standard account of *jus ad bellum*, applying it to such symmetrical
cases as the world wars and "Iraq #1." We pondered, in particular,
the deep reasons offered for there being a just cause to go to war,
especially as that provides a clear, core contrast to both realism and
pacifism. Second, we examined in detail the rise of so-called revisionist JWT, and how this applies to crucial contemporary cases of
asymmetrical war, notably AHI/R2P, the GWOT, "Iraq #2," and
a number of civil war contexts. Ever present, we discovered, was
the issue of political legitimacy, as the right to govern and the right
to go to war are profoundly entwined.

5

Just War Theory and International Law: Conduct-during-War

After armed conflict has begun, how are belligerents supposed to fight? Pacifists, we've seen, insist that things should've never come this far and, at most, only tactics of organized non-violent resistance are permitted. We canvassed the pros and cons of such in chapter 3. Realists, by contrast, say that, once war has begun, it depends on the situation. In principle, anything goes and everything is permitted; in practice, it may be wise to restrain oneself. It depends on the nature of the enemy and what they are after, whether they are a great or small power, one's own military capability, and what's most likely to contribute to one's objective in light of the above. Chapter 2 sussed out the complexities.

Both just war theory (JWT) and the laws of armed conflict (LOAC), by contrast, levy a rather firm set of permissions and prohibitions regarding what belligerents may do during armed conflict. In the first instance, as mentioned in the last chapter, the LOAC rules were negotiated by, and are applicable to, the national militaries of state governments engaged in symmetrical war. So, as with the last chapter, we'll begin our exploration of such rules with classic cases like that in mind, and then move on to consider how non-state actors factor into the *jus in bello* (i.e., "justice-in-war") equation, alongside various contemporary forms of asymmetric armed conflict, notably the global war on terror (GWOT). As with the last chapter, we'll be especially interested to see how JWT offers explanations and justifications for the LOAC, as well as noting where and why they can sometimes come apart. It's a tight, complex interaction between them, with plenty of mutual support and like-mindedness but also some gaps and a few differences, particularly in light of revisionist JWT.

Whereas responsibility for adhering to the *jus ad bellum* rests with political leaders possessing "the war power," responsibility for ensuring adherence to the *jus in bello* falls upon the shoulders of those military commanders, officers, and ordinary soldiers who actually deploy force and do the fighting. (If they violate these forthcoming rules, they can find themselves charged with war crimes after the conflict, either domestically through their own military justice system or internationally.) This clean split of responsibility is the traditional view, and is manifestly the case with the LOAC, but we'll see how JWT revisionists think that this may not, from a moral perspective, tell the whole story.[1]

5.1 The Function of the *Jus in Bello*

The purpose of the *jus ad bellum*, we saw, is to try to ensure that wars are only begun for defensible reasons, such as to protect from rights-violating aggression. The overall function of the *jus in bello*, by contrast, is two-fold: (1) to uphold a kind of ideal regarding a fair and decent way to fight; and (2) to put clear limits on the amount and kind of force being deployed, in particular so that so-called "total war" – i.e., unlimited, indiscriminate escalation in violence – does not happen.[2]

The first function often sounds odd, to both pacifists and realists. To realists, because there's only "smart and strategic" fighting, and *never* "fair and decent" fighting: that would be to make *a category mistake* between self-interested prudence and idealized or even altruistic morality. And what a silly, perhaps very dangerous, mistake it can be: like agreeing to fight with "one arm tied behind your back," when your enemy is literally trying to kill you. One fights to prevail, or one gets eliminated.[3] A quick anecdote captures the alleged oddness from the pacifist perspective: I was once at a conference filled with international law types, and there was much "insider" LOAC talk about permissible killing, collateral damage, and other aspects of the *jus in bello*. In the midst there was a pacifist professor from Australia whose face was turning ever more red. I thought steam might shoot out of his ears. Soon, he could contain himself no longer, and he suddenly shouted: "You mean those rules we're supposed to follow ... as we go about trying to kill each other?!!" This was an effective way of making his point. Yet the answer, admittedly, is indeed: "Yes."

The notion from the JWT/LOAC perspective is this: it may well be sad and bad that there's going to be a fight. But that's what's going to happen. Should we just let anything occur: Any kind of weapon? Any kind of tactic? Targeting anybody? Or, rather, should we insist that, though the whole thing may be regrettable, even in the midst of such fighting there should still be some limits and restrictions observed so that things don't get completely out of hand, and horrifying things like civilian massacre, ethnic cleansing, genocide, or the use of WMD occur? *The analogy is to something like boxing*: there's going to be a fight, but we cannot just let the fighters do whatever they want. If one boxer fights the proper way, whereas his opponent has taken steroids, punches anywhere, and moreover has poison-tipped, razor-sharp barbed wire concealed within his gloves, no sensible person thinks that's alright. There's a difference between *fighting well* versus *fighting dirty*, between fighting with honor and according to some principles, versus fighting with treachery and savagery, perhaps with nasty, hidden implements which are cruel, cowardly, and unfair. The JWT/LOAC idea is that this analogy makes perfect sense, and it's both smart and moral to try to reign in even something as destructive as warfare, to try to ensure that the kind and level of violence does not become barbaric. Michael Ignatieff refers to "the warrior's honor" in this connection, as Shannon French does to "the code of the warrior."[4]

The *jus in bello* is actually much older than the *jus ad bellum*, and much more detailed. Cynics might say that this is because political leaders – who negotiate the international treaties that become international law – have generally found it much easier to impose restraints on their militaries than upon themselves. In any event, there have been dozens, even hundreds of treaties specifying various rules of right conduct during battle. Perhaps the two most famous are The Hague Conventions (referring to 11 separate treaties crafted and ratified between 1899 and 1907) and The Geneva Conventions (of 1949, with additional "protocols" in 1977 and 1996). I'll mention other treaties as relevant but, for now, what we need to know is that, in spite of the thousands of specific rules and regulations which these hundreds of LOAC treaties spell out, all together they serve to realize six general principles of just war fighting.[5] They are:

- discrimination and non-combatant immunity (NCI)
- benevolent quarantine for prisoners of war
- proportionality
- no use of prohibited weapons
- no use of means "*mala in se*"
- no reprisals

Let's investigate them.

5.2 The Six Rules of Right Conduct in War

5.2.1 Discrimination and Non-Combatant Immunity (NCI)

This is the most complex of the *jus in bello* rules, containing several parts. It also raises the most questions, with the deepest implications. Discrimination here means the need for fighters to distinguish, or discriminate, between legitimate and illegitimate targets, and to take aim with armed force only at the former. A legitimate target is anyone, or anything, that is a material part of "the war machine" of the enemy society. (The war machine refers to the military–industrial–political complex that directly guides the war and fights it.) More theoretically, *a legitimate target is anything which is a source of direct, deliberate, and serious physical harm to oneself.* The clearest case is that of enemy combatant soldiers: they have weapons; they've been trained to kill; and they're trying to harm you. Thus, you may treat them with hostility. Specific examples of legitimate targets during armed conflict include:

- soldiers, sailors, marines, pilots, and their officers
- their weapons and equipment
- their barracks and training areas
- their means of transportation
- their supply and communications lines
- the industrial sites where their supplies, weapons, etc. are stored or produced.

Core political and bureaucratic institutions are also legitimate objects of attack, in particular the enemy country's defense department or ministry of national defense.[6]

Illegitimate targets include:

- residential areas
- schools
- hospitals
- farms
- churches/sites of religious worship
- cultural institutions
- non-military industrial sites.

In general, anyone or anything not demonstrably engaged in military supply or military activity is out of bounds and should be immune from direct, intentional attack. Thus, non-combatants (i.e. unarmed civilians) are considered immune from intentional attack. After all, most civilians do *not* have weapons, they have *not* been trained to kill, and they are *not* coming after you with the objective of causing you serious physical harm. They are thus not, to use a term of art frequent in the literature, "dangerous men." It's for this reason that the intentional killing of civilians is seen as probably the worst, and most obvious, war crime.[7] Consider the infamous case of the My Lai Massacre.

Capitalist America fought the Communist Viet Cong during the Vietnam War (circa 1954–75), one of the largest and lengthiest armed conflicts that occurred during the Cold War. The Viet Cong refused to fight the Americans in the open, lest they be blown away by superior US firepower. Instead, the Viet Cong used guerrilla tactics, as defined in chapter 1: hiding, and then striking quickly and unpredictably, before fading back either into the jungle or into small villages, where they would then blend in with the local population, much to the frustration of American soldiers and strategists.[8]

On March 16, 1968, a unit of the US Army commanded by William Calley entered the village of My Lai and rounded up everyone – man, woman, and child. When the villagers refused to identify the Viet Cong soldiers in their midst, Calley snapped and ordered the immediate and systematic execution of everyone. Between 350 and 500 people died, the vast majority of them thought to be completely innocent. Most were women, children, and elderly citizens. (It was later determined that there were, perhaps, 20 or so genuine Viet Cong hiding in their midst.) Calley – but only Calley – was later tried and convicted of war crimes by the American military justice system. But even Calley was later, for various reasons, pardoned by then-President Nixon.[9]

Bin Laden's Objection: Civilians Finance the Harm Which Soldiers Do
Terrorism will always be defined by JWT/LOAC as unjust and illegal, as it violates this foremost *jus in bello* rule of NCI. Terrorism's whole agenda is precisely to target civilians deliberately with death and destruction, believing that such will spread fear and "terror" throughout a population, which in turn is hoped to advance some political goal which the terrorist has. So, for instance, 9/11 was motivated to spread terror throughout the American people, such that they might pressure the US government to withdraw from the Middle East, leaving Osama Bin Laden and other extremists with much more room to execute their agenda (such as, e.g., overthrowing the government of Saudi Arabia, Islam's holy land, and

installing there an extremist Islamic regime or "caliphate").[10] Bin Laden tried to justify his ends and means not merely on his own private politico-religious ideology. He also made an argument relevant to JWT/LOAC, maintaining that civilians are part of the war machine, insofar as the military needs their tax revenues to do what it does. (Furthermore, many militaries, especially in democracies, draw their enlistment ranks out of the civilian population.) Since civilians finance the harm which soldiers do, this means – he once declared – that *everyone* in the enemy society is, in fact, a legitimate target, including for his terrorist schemes.[11]

JWT/LOAC deny this argument, giving several reasons for doing so. First, taxes are coercive in every country: you must pay them, or else you pay heavy fines and/or get jailed. Usually, we think that culpability for something only follows when one has had a free choice to do the thing in question. Second, there's a very indirect, distanced relationship indeed between the act of paying one's taxes and the execution of military action, with a great many intervening steps in-between. Indeed, *it's the state government which is the main agent by far*, in terms of taking the taxes and then translating them into resources for the military, which it may subsequently order into action, sometimes on the other side of the world. The government is so much so the main agent, in fact, that JWT/LOAC submit that it's the only true and fair target: it's the war machine, or nothing. As Thomas Nagel explains: "[H]ostile treatment of any person must be justified in terms of something *about that person* [Nagel's italics] which makes the treatment appropriate."[12] Since the relationship in question is so indirect and distant, and since civilians are not "dangerous men" as described above, it stretches all plausibility to assert they're legitimate targets in war. How may *unarmed* people be targeted with *armed* force? There's a basic failure of logic and proportionality there. Besides, there will still be those in civilian society who fail even Bin Laden's own sweeping criterion: young children, for example, don't pay any taxes, and in any event are truly below the age of having any kind of moral responsibility attributed to them. Thus, Bin Laden's blanket notion that "everyone's fair game" fails to persuade. Moreover, such an attitude would precisely lead to the very thing – absolute, total, indiscriminate, "anything goes" warfare – which the *jus in bello* is designed to prevent.[13]

Dual Use Targets and "Shock-and-Awe"

Civilians are supposed to be off-limits, but what about so-called dual-use targets? These are targets – mostly pieces of infrastructure – used *both* by the military and by civilians during war: roads, bridges, radio and television networks and transmitters, railway lines, harbors, airports, and so on. JWT/LOAC generally forbids these to be targeted but, in reality, they often are, as they're vitally important in helping military planners communicate with their troops and to move them around to where they can fight. More controversial is targeting basic infrastructure, like farms, food supply, sewers, water treatment plants, irrigation systems, water pipelines, oil and gas pipelines, electricity generators, and power and telephone lines/towers. The civilian population pays a huge price for any damage inflicted on such vital social infrastructure, and so it seems to violate civilian immunity to make them targets. The United States has done this twice recently. During the opening days of both the 1999 Kosovo War, and the 2003 Iraq attack, the US conducted a "shock and awe" campaign, relying on air power, bombing raids, and cruise missiles to inflict heavy damage on basic infrastructure (especially communications and electricity lines) in Serbia and Baghdad, respectively. The military goal of such a strike, which is redolent with realism, is to hit the enemy as fast and furiously as possible, dazing them, and "softening them up" for a subsequent ground invasion by army soldiers. It's also to shock the civilians in that society into putting pressure on their regime to give up and surrender quickly. But from a JWT/LOAC perspective, it's questionable practice at odds with at least the spirit, and possibly the letter, of NCI.[14]

Minimizing Civilian Casualties: Requiring Due Care

One of the bitter ironies of NCI is that, on the one hand, it's the most frequently and forcefully *mentioned* principle in the LOAC while, on the other, it's probably the most frequently *violated* principle as well. For civilian casualties have outnumbered military casualties in every known major war since World War I, often by an enormous margin.[15] How can this grim fact be true if the nations of the world truly care so much about the safety of civilians? A few possibilities to consider:

- Not all regimes care about civilian safety. In particular, non-democratic regimes, which are typically propped up with the help of the military (rather than the popular support of the people), have less reason to care about minimizing civilian casualties.
- Many civilians who die in major wars do not die directly from military attack but, rather, from the indirect consequences of military activity. Many, for instance, die from starvation and illness when food supply routes and basic social infrastructure (e.g. water supplies and hospitals) are destroyed by military activity.
- NCI does not make it illegal for civilians to die in wartime. What is illegal is *taking deliberate and intentional aim* at civilians with armed force, as above at My Lai or as today's terrorists do. If a fighting side has taken every reasonable effort to avoid and minimize civilian casualties, then *the death of some civilians by accidental or indirect means is not a war crime*. Civilian casualties under such circumstances are viewed as "collateral damage" – accidental, unintended victims of the fighting. For example, if during an air-bombing raid on an enemy's industrial sites a few bombs accidentally go astray, hitting a nearby residential area and wounding or killing some civilians, these civilians would be considered collateral damage.[16]

So, civilians are by JWT/LOAC entitled *only* to "due care" from fighters; they are *not* entitled to absolute and fail-safe immunity from warfare. What does due care include? It includes serious and sustained efforts, from the top of the military chain of command down to the bottom, *to protect civilian lives as much as possible under the difficult circumstances of war*. So, for example, strategists must make their plans with an eye to minimizing civilian casualties; intelligence needs to be gathered and analyzed to determine which targets are permissible; soldiers need to be trained exhaustively in proper (i.e. restrained, discriminating) ways of fighting; soldiers may in general need to take some added risks upon themselves to ensure they hit only legitimate targets; certain weapons should be favored over others (e.g., so-called "smart bombs"); civilians may merit advance warning of attack, depending on the circumstances; and certainly any rough treatment of civilians needs to be investigated and punished.[17]

Though intended to be assuring, and adding many precautions, all this talk of "due care" is something that pacifists find disturbing. "Due care," to them, seems thin protection indeed compared to genuine immunity, which ironically is in the very name of this JWT/LOAC rule. It takes us to the heart of the objection by deontological pacifism as described in the last chapter. Recall that argument was there made, by Robert Holmes, that one of the most basic and important moral duties is not to kill innocent human beings, which we've seen is how JWT/LOAC define civilian non-combatants: they're not "dangerous men," they've done nothing to deserve death. Yet, Holmes notes, JWT/LOAC permit their killing in wartime, provided some cautions are observed! For Holmes, this reveals the profound ethical problem with JWT/LOAC, and why pacifism must be seen as morally superior. The JWT response involves something called the Doctrine of Double Effect (DDE).[18]

The Doctrine of Double Effect: Revisionism and Replying to Deontological Pacifism

The DDE is a moral concept, invented by Aquinas, which stipulates that an agent A may perform an action X, even though A foresees that X will result in *both* good (G) and bad (B) effects, *provided all* of the following criteria are met: (1) X is an otherwise permissible action; (2) A only intends G and not B; (3) B is not a means to G; and (4) the goodness of G is worth, or is proportionately greater than, the badness of B.[19] Assume now that A is an army and X is an otherwise permissible act of war as defined by JWT/LOAC, like taking aim at a legitimate military target, such as a weapons factory. The good effect G would be destroying the target, the bad effect

B any collateral civilian casualties. The DDE stipulates that A may still proceed to target the factory, provided that: A only intends to destroy the target and not to kill any civilians; that A is not using the civilian casualties as means to destroy the target; and that the importance of hitting the target is worth the collateral dead.[20]

We've seen what the first condition means, in a just war context: the action must be otherwise consistent with the *jus in bello*. What about the second, surrounding "intention"? The JWT/LOAC perspective is, as Michael Walzer says, that *we know the intentions of agents through their actions*: "[T]he surest sign of good intentions in war is restraint in its conduct."[21] In other words, when armies fight in strict adherence to *jus in bello* – taking aim only at legitimate targets, using only proportionate force, not employing intrinsically heinous means – they cannot meaningfully be said to intend (i.e., to want or wish for) the deaths of civilians killed collaterally.[22]

The DDE's third condition is thought to be straightforward, even empirically discoverable: are you trying to use civilian casualties as leverage against the enemy's war machine? This would be revealed, say, by a clear and large ongoing pattern of civilian targeting. Or, perhaps the clearest illustration – apart from terrorism – would be literally taking civilians as hostage and holding them until one gets some kind of gain or concession from the enemy. But when and if, instead, you're taking all the due care precautions mentioned above, and there's no large and clear pattern of ongoing civilian bombardment but, rather, only isolated and small numbers of such, then one can't plausibly be said to be leveraging them against the war machine, and so the third condition would be met. (Perhaps an indirect historical illustration would be the atomic bombings of Hiroshima and Nagasaki in 1945: it's tempting to think that, not only was this deliberate targeting of civilian population centers, violating the first aspect of the DDE, moreover this was done to pressure the military and political establishment in Japan to do what the US wanted, namely, sign a surrender which was unconditional. If that's accurate, then the conclusion is that the atomic bombings could not be justified by appealing to the DDE, as both the first and third conditions weren't met.[23])

The truly difficult aspect of the DDE, and the nub of the dispute between deontological pacifism and JWT/LOAC, is the fourth criterion: contending that the goodness of hitting the legitimate military target is "worth," or proportional to, the badness of the collateral civilian casualties. A pacifist, for example, will always deny this, and assert that civilians are owed true, absolute, fail-safe immunity, and JWT/LOAC – alongside realism and actual military practice – cannot deliver this, and so are all unjust. Pacifism stands alone as the only ethically defensible option. The best JWT/LOAC response is going to involve two things: (1) a rejection of pacifism; but also (2) a preference for the revisionist form of JWT over the more traditional form. Why the second? Precisely because it's needed to respond, in a satisfying way, to this potent challenge from deontological pacifism. After all, how can it be morally justified to foreseeably kill innocent civilians in order to hit a target which only serves the final end of an aggressive war, as defined in the last chapter? The *only* justification sufficient, at least in my mind, to justify collateral civilian casualties (assuming, importantly, that all due care has been satisfied) would be that the target is materially connected to victory in an otherwise *just* war. This suggests that aggressors not only violate *jus ad bellum*, but *in so doing* face grave difficulties meeting the requirements of *jus in bello* as well. To be as clear as possible: to satisfy the *jus in bello* requirement of discrimination/NCI, a country when fighting must satisfy all elements of the DDE. But it seems that only a country fighting a just war can fulfill the proportionality requirement in the DDE: our sense of "worth-it-ness" *must refer* to the overall objective of the war, especially when we're talking about something as serious as civilian casualties. Thus, an aggressor nation fighting an unjust war – for that very reason – also violates the rules of right conduct. This has the important implication that traditional insistences on the separateness of *jus ad bellum* and *jus in bello* may not be sustainable, and that the revisionist account of JWT (in addition to enabling discussion and evaluation of current forms of asymmetric conflict, revealed last chapter) may also be required to deal with this most pointed pacifist challenge.[24]

Note, though, that the "dealing with" shouldn't be taken as settling the issue definitively, merely highlighting the core, logical, doctrinal difference of opinion and value between pacifism and JWT/LOAC on this central matter. For the deontological pacifist, there must be a consistency between means and ends in moral theory and, for them, the means of fighting as permitted by JWT/LOAC are corrupt − notably, by allowing for the killing of some people defined as having done nothing to deserve being killed − and can't be redeemed by such noble ends as trying to defeat aggression. (Revisionist) JWT, on the other hand, has the inverted perspective: *given* the overriding importance of doing things like defeating the Nazis and saving the Tutsis, and *given* the presumed failure of pacifist measures to be able to deliver those things (discussed in chapter 3), there's no other responsible moral choice but to allow for those collateral civilian casualties we know get produced within every war, but strive to minimize them with all the requirements of due care as suggested above.

Another way of putting the difference is this: pacifism insists on absolute immunity for civilian life during war, whereas JWT/LOAC must deny this, as it will entail outlawing all warfare. Since some warfare is justified − both in terms of just causes like defeating extreme aggressors and *genocidaires*, as well as in terms of non-violent alternatives failing to stop such extremists − it follows that we can't acknowledge absolute immunity for all civilians, and instead should strive more practically to put in its place a detailed, responsible policy for minimizing civilian casualties. Hence, due care. Harm reduction through due care.[25] Recall that, for the deontological pacifist, it's all and only about first principles − respect for duty − and if there's a violation of that, then the action is declared wrong, period. While that has the appeal of simplicity, it comes at the cost of outlawing all warfare, and then the conundrums and difficulties of achieving successful non-violent resistance to the worst aggressors must be confronted. We saw last chapter that JWT/LOAC, by contrast, has a moral core which *blends* an appeal to first principles (like respect for rights, and the performance of duties) *with a robust concern for* likely consequences. Since, for JWT/LOAC, the consequences of not effectively resisting extreme aggressors are terrible, it follows that the first principles can't be the only thing relevant in such situations, and then deontological pacifism seems too simple by half, and irresponsibly oblivious to expected consequences.[26]

Child Soldiers

Last section saw the heart of a complex conflict, both between traditional and revisionist JWT, as well as between JWT/LOAC and pacifism. There's one more tough subject connected with non-combatant immunity, and that's child soldiers. Child soldiers are not unknown in history – Hitler notoriously conscripted German boys during the final, losing days of WWII, desperate as he was for any kind of soldiery – but the modern concern revolves around Africa. In the early 1980s, rebels in Mozambique fought a civil war against government forces. Outnumbered, the rebels tried to even the odds by kidnapping boys – some as young as 8! – from their parents, conscripting them into the rebel cause. The rebels would brainwash the boys, train them, abuse them, arm them, and then unleash them on government forces. Since that time, child soldiers have been used extensively in conflicts throughout Africa (Asia, too). One of the prominent cases involves the Ugandan rebel group The Lord's Resistance Army (LRA), led by Joseph Kony. Since 1986, the LRA has abducted more than 60,000 children and forced them to fight against the Ugandan government. It's estimated that more than 250,000 child soldiers have been used in Africa alone since 1980.[27]

The *jus in bello* conundrum is that, on the one hand, child soldiers *are* armed combatants; on the other, they are *children* and, as anyone with experience knows, an eight-year-old hasn't yet reached the age of moral responsibility. Plus, there's all this dreadful background coercion. The phenomenon of child soldiers drives home the potential profound tragedy lurking around many corners of armed conflict. Here's what JWT/LOAC has decided upon: child soldiers, *as soldiers*, remain legitimate targets in wartime, but the use of child soldiers (less than 15 years old) is now recognized as a war crime, punishable against those adult leaders brutal enough to use them. (Kony, for example, has been indicted for war crimes by the International Criminal Court (ICC) at The Hague: see the next chapter for more on the ICC, but know that Kony, as of writing in 2018, still remains at large.) There are also necessary, encouraging efforts by such NGOs as Save the Children to help rehabilitate former child soldiers and ease them back into productive life once the war has ended or they have been liberated from the fighting.[28]

5.2.2 POWs, Benevolent Quarantine, and Torturing Terrorists

It follows from the idea of non-combatant immunity that, should enemy soldiers *cease* being sources of harm during war (e.g., by laying down their weapons and surrendering), then they may *not* be targeted with lethal force after that point. In fact, enemy soldiers who surrender are to become prisoners-of-war (or POWs): they are to be offered "benevolent quarantine" for the duration of the war, and then returned to their home country in exchange for one's own POWs. This is the second general principle of *jus in bello*. Benevolent quarantine means that captured enemy soldiers can be stripped of their weapons, incarcerated with their fellows, and questioned verbally for information. But they can't be tortured during questioning, nor can they be beaten, starved, or somehow medically experimented on. They can't be used as human shields between one's own fighters and the opposing side; the understanding is that POWs are to be incarcerated far away from the front lines of combat. Very basic medical and hygienic treatment is supposed to be offered – things like pain medication, soap, water, and toothbrushes – and, while making captives engage in work projects is permitted, the Geneva Conventions actually require that, under those circumstances, captives be paid a modest salary for their labor. This last condition is rarely met in the real world, but it's quite common for belligerents to disarm, house, and feed their captives, keeping them out of harm's way and ensuring their basic needs are met until the war ends. After all, one hopes one's own POWs are being treated as well by the other side.[29]

Two points remain controversial here:

1. At what point does aggressive questioning become a form of torture?
2. Do non-state actors (such as terrorists) who are taken prisoner deserve the same quality of treatment as state captives?

There's a sense that a soldier fighting for his community deserves better treatment than a terrorist fighting for his cause. This distinction can be difficult to sustain, though, and courts worldwide have found that, generally, non-state actors captured during conflict should be accorded approximately the same rights as captured enemy soldiers. Yet, it needs to be acknowledged that the Geneva Conventions – which mainly deal with the treatment of POWs – were negotiated and ratified by state governments, and they only talk about protections for members of state militaries. The literal reading of them *excludes* non-state actors, and so subsequent court cases have had to "read them into" the treaties, so to speak, based on the law's drive toward universality and consistent treatment and the enduring observation that, once disarmed and in custody, none of these people are "dangerous men" anymore.[30]

This topic was recently ignited during America's post-9/11 round-up, detainment, and treatment of suspected terrorists, especially in Guantanamo Bay, Cuba, and the Abu-Ghraib prison in Iraq during the US occupation of that country (more next chapter). The US Senate Select Committee on Intelligence, in late 2014, released a report on the "enhanced interrogation techniques" used (mainly by the CIA) on people at these installations, especially from 2002 to 2009, and concluded that many constituted forms of torture as prohibited by the Geneva Conventions: prolonged sleep deprivation; prolonged deprivation of food and drink (followed by "forced rehydration" through various body cavities); slapping and punching; waterboarding (i.e., a drowning-based "questioning session"); exposure to deafening noise and extremes of temperature; binding prisoners' limbs so tightly, and for such long periods of time, that sprains, joint separations, and/or bone breakages would occur; death threats to the prisoners and their families; urging or allowing attack dogs either to threaten, or even to bite, the prisoners; and more. Not only do such interrogation techniques violate the Geneva Conventions – though no US official was ever charged with a war crime in this connection – another crucial finding of the US Senate was that *no worthwhile intelligence* (e.g., on any future terrorist attack) *was actually gathered from questioning sessions that made use of such techniques.* This is vitally significant, as sometimes torture is portrayed as a necessary tool for gathering evidence: these findings powerfully suggest otherwise. (Indeed, apparently getting access to their smartphones and social media accounts, or even just following such suspects around as they go about their business, yields much better information than torture. The pain and terror created by torture interferes with accurate recall, and the overriding objective of the victim suspect becomes simply to get the torture to stop, and so they say anything they think will please the torturer enough to do so, including lies and disinformation.) The report, written by Senators from both parties (with some of the very highest levels of access to classified information), paints a very disturbing picture indeed of a dark period in the GWOT. And it may interest people to know that, in spite of promises by various presidents to do otherwise, as of writing in 2018, there are still more than 40 people detained at the Guantanamo prison.[31]

While questioning is permitted under the Geneva Conventions, the inflicting of physical harm cannot be. Why? Because it's impossible to square the inflicting of *physical harm* with the concept of *benevolent quarantine*. Benevolent quarantine may not mean actually being nice to your prisoners, but it cannot, logically, include techniques that the Geneva Conventions define as torture. In addition to the techniques listed above, these also include: strangling, the breaking or severing of limbs or digits, shooting, any kind of electrocution-based session, sexual assault or rape, and poisoning or medical experimentation. These things are simply prohibited. In domestic society, for instance, we don't permit prison guards to torture *anyone* – even those convicted of the very *worst* crimes (much less mere suspects). So, why should we allow it in international society? The answer is: we shouldn't, and we don't. The ban on torture provides a gripping, graphic illustration of the overall function of the *jus in bello*: there are some things you're simply not allowed to do (not even in wartime). You have to find other ways, better ways, to win.[32]

5.2.3 *Proportionality*

The *jus in bello* version of *proportionality* mandates that soldiers deploy only proportionate force against legitimate targets. The rule is *not* about the war as a whole; it's about tactics *within* the war. Be sure, the rule commands, that the destruction needed to carry out a goal is proportional to the good of achieving it. The crude version of this rule is: "Don't squash a squirrel with a tank, or shoot fly with cannon." *Use force appropriate to the target.*[33]

5.2.4 No Use of Prohibited Weapons

Aside from The Hague and Geneva Conventions, there are many treaties which prohibit weapons, such as those banning the use of chemical weapons (1925), biological weapons (1972), and weapons which are "excessively injurious" and inflict "wanton and superfluous suffering" (1980). (This last category includes, for instance, flamethrowers, bullets which expand upon impact, or grenades with nails in them.) Also relevant are the conventions against genocide (1948), against "methods of warfare which alter the natural environment" (1977), and banning land mines (1997). Prohibiting certain categories of weapons puts an added restriction on belligerents and, as such, is consistent with the deepest aim of *jus in bello*, namely, to limit war's destruction. Furthermore, such weapons have, historically, been "taboo." And the very notion of "taboo weapons" or "taboo tactics" reminds us of the concept of military honor equally integral to the *jus in bello*.[34]

Weapons of mass destruction (WMDs) fall under the category of prohibited weapons. WMDs are capable of generating unusually large casualties and property destruction. They are also known as NBC weapons: *nuclear, biological,* or *chemical.* Yet nuclear weapons, as a category unto itself, have not yet been outlawed by treaty, because the major nuclear powers have blocked any such move (even as they move to prevent or deter other nations – especially rogue states like Iran – from acquiring or developing their own nuclear weapons). Nuclear weapons unleash an atomic explosion, causing devastation to physical structures and radiation poisoning in people. They are, to date, the most destructive weapons yet invented. They've been used in battle just twice in all of world history, and it's worth noting that the current generation of nuclear weapons is very much more powerful than those that destroyed

Hiroshima and Nagasaki in August 1945. The list of countries that have admitted to, or are suspected of, possessing nuclear weapons includes the following: Britain, China, France, India, Israel, North Korea, Pakistan, Russia, and the United States. Germany and Japan, together with such wealthy middle-power countries as Australia and Canada, are thought to be capable of building such weapons easily, but have decided not to, for reasons of history or conviction. The move to prevent the spread of weaponry throughout the international community is known as non-proliferation.[35]

Both chemical and biological weapons *have* been outlawed by treaty, but research on chemical and biological weaponry, and stockpiling these "for purely defensive and deterrent purposes," are either permitted or at least allowed to go unpunished. Dozens of nations have, or could have, these devastating weapons. Chemical weapons unleash a gas, or some other chemical, that kills or harms those exposed to it. Mustard gas, for example, was used by both sides during World War I (1914–18) and the Iran–Iraq War (1980–89). In 2012–14, the United Nations (UN) confirmed that the Syrian government, in the midst of the brutal civil war there, used mustard gas, chlorine "barrel bombs," and sarin nerve gas on the opposition: all war crimes. (Indeed, credible reports of such usage have continued up until time of writing in 2018.) Biological weapons release a living organism – usually a virus or bacteria – capable of harming or killing those exposed to it. These are widely dreaded for their unpredictable side effects, but this hasn't stopped biological weapons from being used in battle – such as by the Japanese against China in the late 1930s. Some say the oldest kind of WMD is to surround a town/city militarily, thus putting it under siege, and then trying to poison its water supply (e.g., by throwing in garbage or rotting animal or human carcasses): records on such go back to the very beginnings of warfare.[36]

5.2.5 No Use of Means "Mala in Se"

There's a traditional (yet, not treaty-based) ban on means *mala in se*, or "methods evil/bad in themselves." The imprecise yet interesting JWT idea here is that some weapons, means, and tactics of war are

forbidden not so much because of the badness of the *consequences*
they inflict but because they themselves are *intrinsically awful*. Using
rape as a tool of warfare – for instance, to drive a population off
a territory, or to "reward" one's troops after battle – is a clear
example. Rape is ruled out here not so much because of all the
pain it produces, or because it's aimed at civilians, but because the
act itself is rights-violating, a complete disregard for the humanity
of the woman raped: a coercive violation of her bodily integrity
and her entitlement to choose her own sex partner(s).[37]

Rape has recently been used as a prominent tool of war both
in the Bosnian Civil War (1991–5) and in the wars in the Congo
area of Africa from 1994 to 2008. It was used as an instrument of
terror, as an assertion of dominance, and even out of bizarre beliefs
about trying to "dilute an enemy's bloodline" via inserting one's
own into it. In any event, the extent and consequences of mass
rape in wartime are only now becoming more fully understood,
and the disturbing phenomenon reminds us of how deeply *gendered*
armed conflict can be – i.e., men and women having profoundly
different kinds of experience – which we canvassed in chapter 1
while discussing the feminist analysis of war.[38]

Campaigns of genocide, ethnic cleansing, and torture also fall
under this category of means *mala in se*. In the future, perhaps the
use of child soldiers will be considered such. One of the original
senses of means *mala in se* was forcing captured POWs to fight
against their own side: simply wrong in principle, and at odds with
military honor, regardless of any cost–benefit calculation.[39]

5.2.6 No Reprisals

A reprisal would be this. Let's say that Country A violates one of
the above principles of *jus in bello*: for example, it uses a prohibited
weapon (like chemical gas). May Country B, in response, violate
the same, or a different, rule of *jus in bello*? Reprisals are forbidden
in the LOAC, and the JWT grounds for such are that reprisals are:
(a) a breaking of the rules which are supposed to be observed; and
(b) a recipe for escalation toward indiscriminate and total warfare.
(It's sometimes said that perhaps reprisals might be ways to punish,

or enforce observance upon, law-breakers. However, given that A was already willing to break *jus in bello*, it's hard to see how B's reprisal would suddenly "scare A straight": the much more likely response is A's counter-reprisal, and thus escalation.[40]) The reasoning here is parallel to that on torture: such things straightforwardly break the rules, and are simply forbidden. One is only entitled to pursue one's objectives *within* "the rules of the game," so to speak and as in accord with the boxing analogy. The desire for revenge doesn't justify breaking the rules, though of course it makes it psychologically tempting. Consider that reprisals have been known to happen in wartime, as when the Allies, as demanded by the British, intentionally fire-bombed the German city of Dresden in 1945, killing about 25,000 people (mainly civilians), as a delayed yet direct reprisal for Hitler's terror bombing of civilian centers in London and Coventry in the UK in 1940, thought to have killed about 40,000 civilians. Nowadays, such reprisal activity is completely forbidden. One might, for example, step up the scope and intensity of one's own *permissible* activities in response to a *jus in bello* violation, by way of sending a message and trying to enforce the rules, but one may not violate the rules oneself. To adapt a phrase: *winning well is the best revenge.*[41]

5.3 Reprisal for "Supreme Emergency"? Justification versus Excuse

Walzer has pursued a version of this issue farther, proposing a so-called "supreme emergency exemption" to the above rules of *jus in bello*. He agrees that, during "ordinary" circumstances of war, reprisals should be prohibited. However, when a community faces a "supreme emergency": i.e., imminent defeat to a brutal aggressor with an agenda of either enslavement or genocide – then such a community should feel free to do *anything, including violating the jus in bello*, to prevent such a horrifying fate. Such would be like the ultimate reprisal against the most severe threat to one's survival as a political community.[42]

This is a complex, controversial proposal, and we stress that this is merely Walzer's personal notion, and the LOAC are clear: *no reprisals, period.*[43] Walzer is probably correct, though, when he says that such supreme emergency situations *can* happen, and perhaps the LOAC have a blind spot in this regard. Surely, if anything counts as a supreme emergency, it's genocide. And genocide has happened.[44] But threats of genocide hook into other issues, like armed humanitarian intervention (AHI) by the international community, and that would obviously need to be considered in a very robust way prior to granting permission to a country to do *anything* it wanted. There are issues of last resort to consider, after all, alongside probability of success. We might say that the R2P doctrine discussed in the last chapter in many ways is supposed to mandate international intervention in a way that would prevent the need for any one community to invoke any kind of "supreme emergency exemption" from *jus in bello* rules. Admittedly, though, perhaps the international community won't intervene – we consider Rwanda in 1994 – and there may thus be a dereliction of duty in connection with R2P. So, the fine print of the exact details of the situation would matter crucially, as does a full and satisfying definition of "supreme emergency."[45]

Different thinkers defend different options here. You've been put into supreme emergency yet no one is coming to help. Some say, you should still adhere to the rules, even if you go down in flames because, as Socrates declared: "It's better to suffer injustice than to commit injustice oneself." That's certainly a principled stance, and perhaps pacifists might endorse it, but it may be far too much to ask of people, as we know realists would say: one's very survival is at stake here. Others, such as Winston Churchill (whom Walzer credits for inventing the concept), may simply say that, in such extraordinary circumstances, one has the right to break the rules, in order to survive the supreme emergency. Walzer even flirts with paradox here, suggesting that it's almost like "a right to do wrong."[46] Another option – of broader relevance to our discussions in this book, such as that of necessity and the Caroline Affair in chapter 2 – concerns the distinction between justification and excuse. To be justified in doing something, like X, means to have a moral entitlement to do X; it means that *one does nothing wrong* when one does X. An excuse, by contrast, is when you *have* done something bad or wrong, but you had a very good excuse for what you did – so much so that you ought to be pardoned, forgiven, or exonerated for it. Such an excuse, e.g. in domestic law, usually involves you: having no meaningful control over your actions (say, owing to a sudden and grievous medical condition, like cardiac arrest or grand mal seizure); or being subjected to severe coercion (like the child soldiers mentioned above, or someone trapped in a violently abusive relationship); or being under genuine and immediate threat of death, and your natural survival instincts kick in.[47]

In connection with supreme emergency, some might opine that the victim community, assuming that it's truly on its own and there will be no AHI from others, doesn't literally have the right to violate the *jus in bello* – it still does wrong if it does so – but the severe pressure of the situation, and the natural impulse to survive and not be extinguished or enslaved, *may provide an excuse for the use of exceptional measures.* The realist would, of course, agree – to say the least – and you might think that the pacifist would be horrified. Perhaps most of them would be; but I've recently pondered that there may actually be "an excuse variant" of pacifism. This would be the view that war is still never – not ever – morally justified but it may actually be excusable, for instance under some conditions which JWT/LOAC specify: notably to beat the Nazis, or save the Tutsis, and assuming that all plausible means of non-violent resistance have failed. The notion would be, at that point, *there's really no other choice but to resort to armed force (or, here, extraordinary measures).* This would provide a way for the pacifists to preserve their moral values, while still showing accommodation to the sorts of cases which realism and JWT/LOAC take as being definitive of the problem with pacifism. Excuse and necessity, as related concepts, may actually open space for the Big Three to find what might actually be something as close to common ground as is logically possible.[48]

5.4 Excuse Again, and the "Moral Equality of Soldiers" Debate

There has erupted in recent years a vigorous, fascinating debate between traditionalist and revisionist JWT, quite apart from all the non-classical cases of asymmetric war detailed in the last chapter. Following up on the conviction that the various JWT categories are interconnected, some revisionist JWTs like David Rodin, Henry Shue, and Jeff McMahan have wondered whether we can truly treat responsibility for violations of *jus ad bellum* and *jus in bello* as separate and as applying to totally different groups of people: political leaders for violations of *jus ad bellum*, and officers and soldiers for violations of *jus in bello*. In particular, the revisionists argue that *we may, perhaps, be justified in holding ordinary soldiers accountable for fighting in unjust wars* (just as we may hold political leaders responsible for *jus in bello* violations, if they involve themselves personally with the details of battle plans).[49] Here is one instance where (revisionist) JWT and the LOAC come apart, as the LOAC is straightforward on this point: ordinary soldiers *cannot* be held responsible for *jus ad bellum* violations, only for those of *jus in bello*. This traditional understanding is referred to as "the moral equality of soldiers" insofar as all belligerent military personnel are held to an equal standard under the LOAC: it's not as though soldiers fighting for a just cause have more entitlements in war-fighting than those fighting for an unjust one. This is thought to be true for at least two reasons: (1) soldiers have only very limited scope for free choice in wartime, focused around their own acts on the battlefield, and certainly *not* in terms of when they're free to go to war in the first place; and (2) enemy soldiers are armed and dangerous men, coming at you, and so if the boxing analogy is compelling, then you're entitled to fight back, whether your country ultimately fights for a just or unjust cause. It would, Walzer swears, be unfair "class legislation" to hold ordinary soldiers, many of whom (in non-democratic regimes especially) are conscripts forced to fight on pain of jail-time or death, to the demands of the *jus ad bellum*. That is the "war power" responsibility of political leadership. Every soldier I've ever spoken with strongly agrees, and invites us (so to speak) to walk a mile in their boots as we consider the issue.[50]

Revisionists, though, counter-argue that, if we're truly serious about eliminating aggression, then wouldn't holding ordinary soldiers *to the duty of refusing to participate in an unjust war* provide a potentially very effective further bar? (We can imagine pacifists nodding their heads enthusiastically.) Perhaps that's what's been missing all along, historically: aggressive leaders can't execute their aggression without soldiers, after all and, if we let soldiers know that they, too, will be held responsible for aggression after war, then if the soldiers *en masse* refuse to participate, we may actually have one of the best and biggest blockers to extreme aggression and mass atrocity crimes. Like supreme emergency, this is a complex debate which has attracted a large and spirited literature.[51] Here, we note the basic clash of values. It raises issues, on the one hand, of fairness to ordinary soldiers caught up in circumstances very difficult for them to control individually with, on the other, large-scale theoretical concerns about connections between *jus ad bellum* and *jus in bello*, and practical concerns about reducing or eliminating aggression by upping the obstacles and disincentives. We note that the concept of excuse may also be relevant to this debate. In particular, that it doesn't have to be an "either/or" choice between holding soldiers responsible versus immunity from such: we may judge that full-blown moral and legal responsibility for *jus ad bellum* most properly lays with the political leaders who set the war in motion, and while perhaps there may be some *ethical* concern or criticism which we might level at some ordinary soldiers in this regard, perhaps they should be excused from *legal* punishment on grounds that: (a) they didn't set the war into motion, they're not in charge of the war machine; and (b) their scope of choice resides overwhelmingly within the *jus in bello*. In other words, it might be ethically ideal if all soldiers were to refuse to fight in unjust wars but, in practice, as individuals they may have very compelling excuses (of coercion, of non-control) for failing to do so, and so legally we leave them out of prosecution for *jus ad bellum* violations. Accountability tracks the scope of choice they truly do have, namely, the *jus in bello* rules as described this chapter.[52]

5.5 Rules for the Rich? An Objection by Some Non-State Actors

Sometimes, people assert that only rich and powerful nations can afford to satisfy the rules of *jus in bello*. These rules are thus said to be unfairly stacked in their favor. Poorer countries, and emphatically non-state actors like terrorists, militias, and secessionists, lacking things like satellite-guided precision weaponry, are thus doubly condemned: condemned (probably) to lose; and condemned as war criminals for using the more "dirty," older, and indiscriminate weapons and tactics within their means. JWT/LOAC completely disagrees. Consider our boxing analogy: we don't think that the relative net worth of the boxers has any kind of impact on how they should fight each other. Similarly, poor nations *can* fight in accord with these rules. It's not too much to ask them to refrain from: bombing, terrorizing, or mass raping civilians; torturing prisoners; dropping down nukes; engaging in reprisals; etc. They might resent the wealth and weaponry of rich nations, *but these emotions don't give them a "poor man's right" to fight dirty*. We know from history, after all, that wealth and advanced weaponry are no guarantee of decent behavior on the battlefield. Indeed, aggression is usually committed by powerful nations against the weak. Fundamentally, whether rich or poor, a group *makes a decision* to fight decently or dirtily. Fighting decently is, indeed, no guarantee of victory (and perhaps that disturbs realists). But there's never any guarantee that doing the right thing is always going to serve one's interests, or make one happy. Think of personal cases where you do your duty (e.g., avoid stealing) even though it prevents you from doing something which would make you happier (e.g., getting a new car without having to pay for it). Poverty, or relative underdevelopment, is no justification for committing war crimes. This complaint, from a JWT/LOAC perspective, amounts to a very bad "excuse" indeed for terrible actions when none is justly warranted. And it's bad enough when made by poorer countries who may wish to otherwise fight decently. But it's quite disingenuous when made by such non-state actors as terrorist groups who are, essentially, trying, as Bin Laden did, to argue that they shouldn't be held to *any* standards at all – everyone's a target, any weapon may be used, anything goes – which should be seen for what it truly is: namely, a total denial of the *jus in bello* and its purpose, and an embrace of classic, unconstrained realism, the nature and weaknesses of which we've already explored in chapter 2.

5.6 Conclusion

We continued our exploration in this chapter of the JWT/LOAC perspective on war and political theory, by examining intensively its understanding of right conduct during battle. The six general principles of proper war conduct were explained, alongside many examples: (1) non-combatant immunity from direct and intentional attack, demanding due care for civilians and as further illuminated by the DDE; (2) benevolent quarantine for captured soldiers; (3) use of proportionate means only against legitimate military targets; (4) no prohibited weapons; (5) no use of means "*mala in se*"; and (6) no reprisals. We considered limits, unclarities, and objections throughout, and at the end focused on three large-scale objections to traditional *jus in bello* as a whole: (1) a proposed "supreme emergency" exemption from the rules; (2) the moral equality of soldiers debate; and (3) complaints that all such rules are biased in favor of more powerful countries. We turn now to the transition from war to peace – the cessation of hostilities and the aftermath of armed conflict.

6

Just War Theory and International Law: End-of-War

Sooner or later, every war comes to an end. Sometimes, this is owing to clear, crushing victory by one side, others because the belligerents simply stop fighting, as they can't see any possibility of winning, and can no longer sustain the expense or the drive to keep on fighting. The termination phase of war is complex and fragile, and all kinds of mistakes and injustices can be made: indeed, these are just as possible during this third and final phase of armed conflict as with the other two of *jus ad bellum* (start of war) and *jus in bello* (conduct during war).

6.1 The Three Big Theories, Historically, on Post-War Justice

Thinking about the cessation of hostilities, and post-war settlements in general, has historically been very underdeveloped, in each of our three major theories. This may surprise with pacifism, as in many ways the whole point of the doctrine is to bring an end to war. Yet, that hoped-for end is *a general end-state*, as opposed to a particular focus on wrapping up *this* armed conflict in a good and decent manner. Pacifists pay overwhelming attention to the start of war, as that's what they are most keen on preventing. Prevention, not termination or closure, soaks up the majority of their efforts. In many ways, pacifists think that everything is lost, at least morally, once war has begun and so the subsequent issues don't interest them, or nearly as much.

The focus for realists, we know, is on winning the war if you can or, if you can't, then at least getting "the least worst" alternative. For most of history, the realist ethos dominated, summed up in the pithy proposition that "to the winner go the spoils of war," and the prevalence of that permissive attitude, alongside the sheer power of those countries most likely to win their wars, translated into a legal and moral vacuum in the aftermath of war. Essentially, and certainly relative to the other two phases of conflict, *the endings of wars were left unregulated*, to be determined by the vagaries of post-war power.[1]

Realists couldn't have cared less about that – perhaps even welcomed it – but you'd think that international lawyers and just war theorists would've cared more. Many, historically, contented themselves with the claim that "well, post-war justice consists in achieving the just cause which triggered the war; and, if that can't be had, the settlement is not going to be just – period." I label this the "mere adjunct thesis": that *jus post bellum* ("justice after war") is merely an adjunct or corollary of *jus ad bellum*, and simply consists in realizing the just cause for which the war began. This was the standard JWT view for the longest time, resulting amongst other things in a belief that there exist only the first two categories of *jus ad bellum* and *jus in bello*, and there's no such thing as a third category of *jus post bellum*. One still encounters such a view sometimes today.[2]

This traditional view seems decreasing in frequency, perhaps emphasizing the decline in traditionalist JWT versus more contemporary, revisionist forms. For, contemporary JWTs point out that *even something as basic as war crimes trials isn't clearly mandated by "the mere adjunct thesis."* Moreover, the traditional view simply ignores all the other issues and questions posed by the aftermath of conflict. Some of these include:

- If a regime committed aggression to start the war, and (even though it may have lost the war) it remains in power at the end, should it be left in power, or forced to give it up? Why or not?
- If, alternatively, such an aggressive regime was destroyed during the course of the war, who's now responsible for re-constituting the government within that society? The local population only? But what if there are big internal divisions amongst them? The war winners? But what if they desperately want to leave – or, rather, engage in self-serving, "neo-imperialist" tinkering? Or should it be the international community in general? A complex kind of partnership amongst all? Yet how would that actually work?
- May we force countries to de-militarize, and/or give up territory, in the aftermath of war? Re-drawing borders can, after all, alter the entire course of history for many peoples.
- What steps should we take to ensure that a peace deal, once agreed to, actually gets observed? How should we handle any back-sliding or disobedience in that regard?

The traditional "mere adjunct" view seems all but empty, and completely unhelpful, with regard to this quick enumeration of potential issues. A much fuller account of post-war justice appears mandated. And not merely for the sake of theoretical comprehensiveness, or greater completeness. The other reason for the huge growth of recent interest in *jus post bellum* involves all the practical, real-world difficulties, over the past 20–30 years, of trying to achieve justice after war, in a smart and stable way, be it in Afghanistan, Bosnia, Colombia, Congo, Croatia, Ethiopia/Eritrea, Iraq, Libya, Rwanda, Serbia, or (South) Sudan. One of the most trenchant truths about warfare, historically, is that *when wars are wrapped up badly, they sow the seeds of future wars*. This is graphically depicted in the relationship, say, between WWI (1914–18) and WWII (1939–45) and between Iraq #1 (1991) and Iraq #2 (2003).

Having an account of *jus post bellum* which goes beyond the "mere adjunct" thesis also allows JWT/LOAC to respond, in a satisfying way, to one of pacifism's most provocative political challenges: that JWT/LOAC are "too complacent" about the set-up of the international system, and do nothing to try to improve things about it, such that wars don't break out as often. A more robust theory of *jus post bellum* offers the possibility of a substantial bridge between JWT/LOAC and pacifism in this regard – and, indeed, to realism, too, seeing as how it must rationally concern itself with the very real threat of bad war settlements igniting second, and costlier, wars. No compelling contemporary account of armed conflict can avoid wrestling with how wars end, with how they should transition from mass political violence into a solid, decent and durable peace. That's the goal of this chapter and, as with our prior examinations of JWT/LOAC, we'll proceed outwards from classic, symmetrical application of these ideas toward present-day asymmetric cases, notably involving Iraq and Afghanistan.

6.2 Drawing on Some International Law

While it's true that, relative to the outbreak and conduct of hostilities, the termination phase is very much underdeveloped, there *is* nevertheless *some* international law in this regard. Parts are quite dated and almost unbelievably quaint: e.g., in Hague IV (1907), Article 32, reference is made, for those who wish to sue for peace (armistice), to march toward enemy lines bearing not only the clichéd white flag but also a drummer and a bugler. It is solemnly declared that such figures should not be shot, as they are "inviolable."[3] Such sweet articles aren't worth much comment nowadays, but there are two large items of LOAC which are substantial and worthwhile: occupation law and war crimes trials.

6.2.1 Occupation Law

Recently, there's been a resurgence of interest in so-called occupation law, owing to high-profile cases directly involving foreign military occupation and rule of a country defeated in war. One thinks immediately of America, post-2001, in both Afghanistan and Iraq, but shouldn't forget the ongoing issue of the Occupied

Territories involving Israel and the Palestinians. Occupation law is controversial, not only for the intrinsically contentious nature of its subject matter, but also on more formal, procedural grounds regarding whether it should still be considered "active and living" international law: whether it was crafted at a time still applicable to our own age; and due to the fact that it refers to various parts of various *other* international treaties, and there's never been one big direct international treaty, widely ratified, exclusively dealing with foreign military occupation. And it's especially difficult to enforce occupation law, seeing as how the occupier (practically by definition) has the vast preponderance of power *in situ*, and the broader international community is often reluctant to, and/or incapable of, effectively challenging that. Even world-class experts on occupation law – notably Eyal Benvinisti, Eric Carlton, and Yoram Dinstein – are divided on these matters. Still, it exists and, as you'll see, has some surprisingly helpful and robust content to it, which would at least be relevant to JWT, even if we wonder about its current legal bindingness.[4]

Should a war winner find itself in charge of a country it has defeated in war, occupation law stipulates that it needs to adhere to at least *the five following general principles* in the aftermath of conflict:

- Any foreign occupation must be temporary. (Otherwise, it's conquest, and you're incorporating that territory and its population into your own political community.)
- The occupier may not engage in collective punishment. (Note the stress on "collective" was, and remains, deliberate, so as not to preclude individualized war crimes trials of culpable persons.)
- The core *jus in bello* principle of discrimination/non-combatant immunity must be respected by occupying forces throughout the period of occupation. (But such forces *do* have the right to defend themselves from armed attack: e.g., by "regime remnants"/left-overs, or an attempted insurgency.)
- Occupation forces must not engage in looting (i.e., widespread, systematic theft and vandalism), and must respect the property of civilians.
- The occupier has a duty to provide for "basic civilian well-

being," especially in connection with: (a) public order and safety; (b) such vital needs as food and water; and (c) such elemental infrastructure, with public health impacts, as water sanitation, garbage collection and sewage treatment.[5]

The method throughout this chapter will be to cobble together components of a compelling revisionist JWT understanding of *jus post bellum*. (This is forced upon us, owing to the underdevelopment of the subject: we can't just report on the existing rules of *jus in bello* and their controversies, like in the last chapter.) Perhaps only at the end will the picture be seen in its entirety, and of course we can't do complete justice to each of the elements here: all the historical cases, the full meanings, all possible dilemmas. Submission is made in this section, for instance, that the five above principles of occupation law are: quite sensible; connect with prior JWT/LOAC values; are directly applicable to even the latest cases of occupation; and have at least one further, essential function. This function is that, in the relevant cases, they explicitly serve as a normative bridge between the in-war period (when *jus in bello* is operational) and the post-war period (when *jus post bellum* predominates). For, as is often noted, there usually isn't a razor-sharp, "night/day" distinction between war and post-war: *the historical norm is much more of a complex transition phase*, often in bumps and starts, between full-scale "hot war" and, eventually, a reliable stoppage of hostilities and the true beginnings of peace.[6]

6.2.2 War Crimes Trials

The need for such trials, from a JWT perspective, follows from Michael Walzer's maxim: "There can be no justice in war if there are not, ultimately, responsible men and women."[7] Individuals who play a prominent role during wartime must be held accountable for their actions and the consequences thereof. This is one of the most important ways *for actually enforcing the laws of war*, crucial for: (a) any sober talk about genuine *laws* of armed conflict; and (b) to reply to sharp realist queries about *the reality* of the JWT/LOAC principles we've explored over the past two chapters.[8]

There are, we've seen, two broad categories of war crime: those which violate *jus ad bellum* and those which violate *jus in bello*. *Jus ad bellum* war crimes have to do with "planning, preparing, initiating, and waging" aggressive war, as defined in chapter 4.[9] Responsibility for the commission of any such crime falls mainly on the shoulders of the political leader(s) of any aggressor regime (though we saw there are some revisionists, like David Rodin, who favor a broader application which might include some ordinary soldiers). Such *jus ad bellum* crimes are, in the language of the Nuremberg prosecutors, "crimes against peace."[10] *Jus in bello* war crimes, by contrast, involve the violation of the six general rules of right conduct in battle, as detailed and analyzed in chapter 5, ranging from non-use of prohibited weapons to non-torture of captured or surrendered enemy soldiers to non-targeting of unarmed civilians. Historically, the vast majority of war crimes prosecutions have been of soldiers and officers for *jus in bello* violations.[11]

Nuremberg (and Tokyo)

The Nuremberg trials refer to those held after World War II, in 1945–6, in the German city of that name. There were Japanese war crimes trials too, in Tokyo.[12] (It's unclear what counts as "the world's first war crimes trial": Germany, post-WWI, was forced to put on trial some of its own soldiers at Leipzig in 1920–2. These "trials" were widely regarded as a half-hearted and ill-structured fiasco, and many historians have opined that the Allies took away the following lesson for WWII: don't let defeated countries try their own, as they won't do it properly.[13]) In Nuremberg, much of what was left of the Nazi hierarchy was put on trial for both war crimes (i.e., starting an aggressive war) and crimes against humanity (i.e., the Holocaust). While every war crime is also a crime against humanity, the converse is not true, since crimes against humanity can be committed in peacetime and cover a broader range of terrible actions, such as grievous human rights violations.[14]

While some objected that the Nuremberg trials were "victor's justice" – realists would certainly agree – it's hard to agree that that's all they were. The process, after all, was rigorous, open, and fair. All those charged had lawyers, could cross-examine witnesses and experts, and present their own evidence. The trials operated on rules known to all. Consider also the raw numbers: 22 Nazi elites were tried over the course of a year, 3 were found innocent, 19 guilty. Of the guilty, 11 were sentenced to death and 8 to prison sentences of varying lengths. This conviction rate, of just over 86%, is actually in line with the average rate of conviction for non-war crimes in most developed societies.[15] Combine this percentage at Nuremberg with the variety of outcomes in sentencing there, and it looks like you have, from a JWT/LOAC perspective, a reasonable judicial process looking at the evidence in each individual's case – as opposed to one which sweepingly condemns and convicts them all because they happened to be on the losing side. (Yet, it must frankly be admitted that no Allied leader, officer or soldier ever faced an international war crimes tribunal, and we know there were very controversial war actions, notably Dresden and Hiroshima/Nagasaki, or the "Katyn" massacre in Poland by the Soviets, and thus it seems sober to judge that *there was at least a "double-standard"* in terms of overall war crimes prosecution, post-WWII.[16])

War Crimes Trials since WWII, Including the ICC
Most national militaries nowadays have their own internal military justice systems, and these can be very well developed. They enforce discipline – punishing things like insubordination, "hazing" (i.e., objectionable "initiation" rituals sometimes performed on new recruits), and sexual harassment within the ranks – as well as alleged war crimes activity during conflict. Recall from the last chapter that Calley, of My Lai infamy, was tried and convicted by the internal US military justice system. I've actually heard US soldiers say, at times, that they are more afraid of US military lawyers – you may have heard of "the JAG corps," short for "Judge Advocate General" – than they are of enemy soldiers. Yet we shouldn't forget the lessons of Leipzig entirely, as we also recall from the last chapter that, for example in connection with the Abu-Ghraib prisoner scandal during the Iraq occupation, only a handful of US persons were ever charged with internal discipline offences, and no

one was ever convicted of a war crime, in spite of all the Geneva Convention violations.[17]

After the brutal civil war which tore apart Yugoslavia in 1991–5, and the Rwandan genocide of 1994, the international community decided to convene war crimes, and crimes against humanity, proceedings. The ICTY (International Criminal Tribunal for the former Yugoslavia) existed from 1993 until 2017, and dozens were tried. While no new trials will take place, a successor institution now exists to handle some outstanding sentencing and appeals items. And the ICTR (International Criminal Tribunal for Rwanda) existed from 1994 until 2015, and there exists today a similar successor institution for it, too. Both of these *ad hoc* – or "one-time" – criminal procedures featured the same commitment to due process as Nuremberg did – perhaps even too much, in the sense that this resulted in a serious backlog in cases, which may have meant that not everyone was tried who ought to have been. The courts also created some political controversies, particularly the ICTY, where accusations of imperfect prosecution across the various belligerent communities generated raw feelings. Still, both courts tried dozens of people complicit in dreadful deeds, and, unlike Nuremberg, were able to try some of the biggest fish: the former prime minister of Rwanda; and the former president of Serbia/Yugoslavia. The results: the former Serb president, Slobodan Milošević, died (of natural causes) during his trial before sentencing could be determined; but Jean Kambanda, former prime minister of Rwanda, pled guilty to genocide charges, and is now serving a life term in prison.[18]

Partly to overcome the limitations of *ad hoc* trials, and to institute something more lasting, there arose in the late 1990s a movement to create a *permanent* international criminal court (or ICC). In 1998, the Treaty of Rome was signed, paving the way for the ICC to be formed, at The Hague in the Netherlands. The goal is to *try all war crimes in all wars, committed by any and every side*. To try to transcend Nuremberg and ensure objectivity – no "victor's justice" nor "double-standards" – those who prosecute, defend, and sit in judgment on the actions in question must be from countries not involved with any of the belligerents in question.[19] After having played a major role in the drive to establish the court, and its rules of procedure, the United States in the end refused to join the ICC, saying that members of its military accused of war crimes will be tried *only* by its own domestic military justice system. This does raise issues related to Leipzig, as noted, but perhaps the more serious issue is that US non-participation may well have doomed the ICC from the start. However, the ICC has started up, and to its credit has managed reasonably well – charging 42 alleged war criminals, with 19 of those cases having concluded (at the time of writing), the rest remaining ongoing – in spite of its numerous challenges:

- not having enough money to pay and keep staff (lawyers, judges, their aides)
- not having enough money to gather evidence and interview/"prep" witnesses
- the fact that war often destroys the very evidence needed to convict
- getting their hands on the accused, as the ICC may only rely on national governments handing over those accused. (Kony, of LRA/Uganda rebel/child soldier infamy described in chapter 5, has been charged and a warrant for his arrest issued by the ICC, yet he remains at large.)[20]

In spite of the strong JWT/LOAC principles behind its existence, the future of the ICC is today somewhat in doubt. US non-involvement, and refusal to offer any resources, is a major blow. History seems to show, and realists wouldn't be surprised, that the support of great powers is essential for the long-term survival and success of an international institution. Furthermore, and at the other end of the power spectrum, there's been a recent backlash by African countries, threatening mass pull-out. Burundi, Gambia, and South Africa have all announced that they intend to exit the Treaty of Rome (though, for example, this has triggered internal constitutional lawsuits within South Africa, which may block its exit). The accusation is that the ICC is targeting African war criminals disproportionately. And, they might say: "Besides, if

America isn't willing to subject its own military service people to the ICC, then why should we?" Indeed, perhaps these thoughts reveal that old concerns about double standards and victor's justice still linger: though we may be impressed with the principles behind the ICC, its diligent efforts in spite of lack of resources, and with the quality of its procedures, the fact remains that those put on trial are, overwhelmingly, those who've lost their wars. It may not strictly be *victor's* justice anymore, as it's neutral international prosecution, but it may be "loser's justice," so to speak, in that only those on losing sides face prosecution. If Assad, for example, winds up winning his civil war in Syria, he may never face trial and may even die still holding onto power, as did Stalin. What does this say about the ideal of trying all war criminals from all wars? We know how realists would reply. For JWT/LOAC, I think the answer must be something like: in practice, it's an admitted work-in-progress; yet, in principle, the commitment to accountability in general must remain rock-solid.[21]

6.3 Drawing on Historical Policies of Post-War Justice

Building on occupation law, plus war crimes trials, what else might be part of a persuasive JWT/LOAC vision of post-war justice? Perhaps historical practice has something to offer. Historically, especially when Western democracies have been involved in war, *and when they have been in a position post-war to implement a vision of justice* (which, of course, isn't always the case, as in Vietnam for example), there tends to be the use either of a policy of retribution or of rehabilitation. Consider the simple Venn diagram shown in figure 6.1. Some intentional overlap is depicted. I'm not sure the overlap is logically mandated, as at heart the two post-war policies differ substantially (as we'll see). But I submit that, whether someone supports retribution more, or whether they support rehabilitation, they also endorse a very small yet vitally important set of five core, "thin," most rudimentary, propositions of post-war justice.

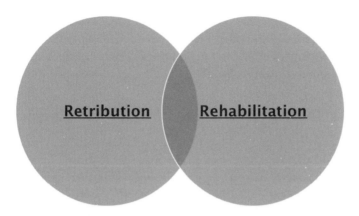

Figure 6.1 Major Policies of Post-War Justice, Historically

6.3.1 *The Thin Theory: Five Overlapping Core Principles of Post-War Justice*

First, there needs to be an actual *ceasefire*. Not much point talking about "post-war" if a "hot war" is still ongoing. A cessation of hostilities, or truce, or armistice, needs to have been arrived at.

Second, there needs to be an *exchange of prisoners-of-war* (POWs). This is also a small way to build some needed mutual confidence regarding a peace deal.

Third, while it doesn't need to be finely detailed, the peace agreement should be *publicly proclaimed* so that everyone's expectations are clear, everyone knows the war is over, and everyone has an idea of what the general framework of the new post-war era will be. (Sometimes, by contrast – such as in medieval Europe – the most crucial parts of a peace treaty were kept secret from the public.[22])

Fourth, all sides must commit firmly to *accountability*. Publicity is part of that, but so too are war crimes trials.

Fifth, there must be *proportionality* between the terms of the peace and the original cause of the war, alongside the actions occurring during the war. So, for example, if a war erupted over where exactly to draw a border between countries, and one side lost, and the whole thing was fairly small-scale and moderate, with all sides generally observing the *jus in bello*, it's going to be disproportionate post-war for the winning side to insist that the losing society "surrender unconditionally" and subject itself to bold, coercive regime change.[23]

6.3.2 The First "Thick" Theory of Post-War Justice: Retribution

Speaking of winning and losing, it's worth reiterating, and explaining, how the two major historical policies or theories of post-war justice are *based on some assumptions and idealizations which don't always hold in the real world*. In particular, they both assume that: (a) there was a clear aggressor, in our chapter 4 sense; and (b) the aggressor has lost the war to the justified victim and/or its allies in the international community. Again, this doesn't always happen in the real world: sometimes, there is neither a clear "bad side" aggressor, nor a clear winner. But proponents of these theories, while admitting the truth of this, will reply that: in such instances, we can't really speak of post-war *justice*. If the aggressor has won, then by definition justice hasn't been served by the war and any post-war "deal" is going to be, at the very least, sub-optimal (if not horribly unjust). So, crucially, it's not as though these two forthcoming "thick theories" are committed to the notion that every war truly does follow either a retributive or rehabilitative template. Quite the contrary. The notion is that: post-war settlements are probably going to be unjust *unless*, at the least, if there *was* a clear aggressor, *then* that aggressor has lost *and* the justified victim and/or its allies are *in a position, and are interested, to implement justice*. The question then becomes: which vision of justice should be implemented? And then we see the split between either a retributive or rehabilitative theory. Lastly, these theories *are themselves ideal types*, thus some may contest what's truly retributive versus rehabilitative: yet historically, the following tends to be implemented.[24]

Retribution's Principles and Rationale

The core intuition behind retribution theory is that *the defeated aggressor must be rendered worse-off than prior to the start of the war*. This is thought necessary for proper punishment, and the essence of justice itself. Why must there be punishment at all? Why can't we just cancel the aggressor's unjust war gains (if there were any), and then "live and let live"? Three reasons are articulated by retribution theorists, such as Robert Nozick. First, the obvious – yet powerful – one of *deterrence*. Punishing past aggression deters future aggression, or at least does so more than if we had no punishment at all. No punishment seems a lax policy which actually invites future aggression. Second, proper punishment can be *an effective spur to atonement and change* on the part of the aggressor (since presumably it doesn't want to suffer through such punitive measures again). Finally, *failing to punish the aggressor degrades and disrespects the worth, status, and suffering of the victim.* Thus, it's not enough merely to take away the unjust gains of the aggressor, and perhaps force it to apologize publicly for its wrong-doing. It must suffer some genuine retribution as a matter of balance and fairness. This is the same reasoning that explains and articulates criminal punishment (*lex talionis*) in domestic society: if someone has stolen something, for example, we don't just make him give it back, and apologize to his victim: we make him suffer further – make him worse-off than prior to his deed – for instance, by making him pay a fine, or spending some time in jail.[25]

How exactly should we make the defeated aggressor worse-off than prior to the war? War crimes trials for those individuals complicit in aggression: absolutely. Next, retributivists say, at least partial demilitarization. Since the aggressor broke international trust by committing aggression, it *can't* be trusted *not* to commit aggression again – certainly not in the short term, and not without a change in government. And the international community, especially the victim, is entitled to the security of knowing there won't be a repeat. Therefore, the tools that the aggressor has to commit aggression must be taken away. Essentially, the defeated aggressor loses many of its military assets and weapons capabilities, and has caps or limits placed on its ability to rebuild its armed forces over time.[26]

Two further penalties are often insisted upon by retributivists: *reparations payments* to the victim(s) of the aggressor; and *sanctions* slapped onto the aggressor as a whole. These are the post-war equivalent of fines levied on the whole society of the aggressor nation. Reparations payments are due, in the first instance, to the countries victimized by the aggressor's aggression, and then to the broader international community. The reparations payments are *backward-looking* in that sense, whereas the sanctions (discussed in chapters 2 and 4) are more *forward-looking* in the sense that they're designed to hurt and curb aggressor's future opportunities for economic growth, at least for a period of time (like "probation") and especially in connection with any goods and services that might enable the aggressor to re-commit aggression. One more element completes the retributive picture: *no aid or assistance is offered to the locals with post-war reconstruction*, and nothing is done to remove the defeated regime from power. Everything is left to the locals, and the war winners are free to leave and wash their hands of it all.[27]

A Recent Case of the Retribution Model
The highest profile case of the retribution model, of course, deals with WWI and the Treaty of Versailles (1919). The Treaty is widely considered a controversial failure which directly contributed to the rise of Hitler and the outbreak of WWII. Since it's so well known, I'll leave it there.[28]

The 1991 terms ending the Persian Gulf War were retributive and likewise paved the way for a second war. The terms called upon Saddam Hussein's Iraq to give up any claims on Kuwait, to apologize for its aggressive invasion of Kuwait in 1990, and to surrender all POWs. Saddam was left in power, no attempt being made either to change his regime or to bring anyone to trial on war crimes charges. But Iraq *was* to be extensively demilitarized: it lost many weapons, and had strict caps put on any rebuilding. It also had no-fly zones (NFZs) imposed on it, both in the north (to protect the minority Kurdish population from Saddam) and in the south (to protect the Shi'a, who make up the majority of Iraq but who were often oppressed, historically, by Saddam's Sunni-led government).[29]

Iraq also had to pay financial reparations to Kuwait for the invasion and, moreover, had to suffer sweeping sanctions on its economy, especially on its ability to sell oil. These sanctions dev-astated Iraqi civilians while doing very little to hurt Saddam. In fact, evidence suggests that the sanctions only *cemented* Saddam's grip on Iraq, as increasingly impoverished citizens grew more and more dependent on favors from Saddam's government in order to survive.[30]

Finally, Saddam had to agree to a rigorous, UN-sponsored process for conducting periodic weapons inspections. The inspections occurred from 1991 to 1998, and uncovered literally tons of illegal weapons (defined in chapter 5), including chemical and biological agents, which were subsequently destroyed. But when Saddam kicked out the inspectors permanently, in 1998, he touched off a major international crisis that contributed directly to war a few years later. Saddam's move effectively confirmed, for many American officials, the view that Iraq still had WMDs and, moreover, was plotting to give some to al-Qaeda to enable another 9/11-style terrorist strike on the US (a view scrutinized in chapter 4). Ostensibly to prevent this, America struck first, in a regime-change war, in Iraq starting March 2003.[31]

Summarizing Retribution's Pros and Cons

Enumerating retribution's advantages: it's simple and straightfor-ward; it's the historical norm (there may be value in long-standing habitual practice); there's no messing around with complex, costly, and controversial regime change and help with post-war recon-struction; and, if one believes that justice requires retribution as a matter of principle, then one is going to be convinced of retribution's claims. This is what the aggressor *deserves*.

Alternatively: if one doesn't believe that justice requires retribu-tion, then retribution theory is going to appear as angry revenge and not balanced justice; the reparations and sanctions called for by retribution theory very often wind up hurting innocent civilians, and thus violate the core JWT/LOAC principle of discrimination/ non-combatant immunity; and, finally, in historical cases where retribution has been employed, there often has been imperfect or even quite shoddy results. The revenge creates bad feelings and new generations of enemies, the economic punishments stoke desperation, and the aggressive regime is left in place, all combining together over time to produce – rather regularly – a second, and much more destructive, war.

6.3.3 The Second "Thick" Theory of Post-War Justice: Rehabilitation

The policy of post-war rehabilitation is, in many ways, motivated by a desire to overcome these negative aspects of retribution policy. While they do still share the five prior principles of the basic "thin" theory, the two "thick" models nevertheless have very different overall goals. Retribution wishes to make the defeated aggressor worse off than prior to the war (so as to punish), whereas rehabilitation strives for the opposite: *to make the defeated aggressor better off than prior to the war.* Why? Fundamentally, out of a conviction that retribution simply doesn't work, regardless of people's emotions or intuitions about "justice"/payback. Indeed, rehabilitators may suppose that: (a) retributivists place too much reliance on the interpersonal analogy to domestic, individual crime; and (b) the broad-based historical evidence shows that *that analogy is misleading, even destructive, at the society-wide level.* In other words: individual war crimes trials are one thing – indeed, are mandated by the thin principle of accountability – but society-wide retribution is quite another.

Toward the overall goal of improving widespread well-being, and in light of the historical evidence, the rehabilitation model *rejects the imposition of both sanctions and reparations.* Not only because these have been shown to hurt innocent civilians, as in Iraq, but, moreover, because post-war reconstruction is difficult enough at the best of times: to succeed *while sucking money out of the target country* becomes all but impossible. In fact, rehabilitation favors *investing in* a defeated aggressor, to help it rebuild and to help smooth over the wounds of war. Finally, the rehabilitation model *favors forcing regime change,* whereas the retribution model had viewed that as too risky and costly, and perhaps as too much of a favor. Indeed, enforcing this condition is costly, but rehabilitators believe it can be worth it over the long term, by enabling the creation of a new, non-aggressive member of the international community. To those who scoff that such deep-rooted transformation simply can't be achieved, rehabilitators reply that, not only *can* it be done, it *has* been done. The two leading examples are West Germany and Japan after World War II.

Two Cases of Rehabilitation Succeeding: West Germany and Japan After WWII

Since these cases are so well known, I move quickly to the abstract "rehabilitation recipe" crafted and implemented by the Allies over the course of their experiences with these two historical situations.[32]

The post-war reconstructions of West Germany and Japan, as guided by the principles set out in table 6.1, easily count as the most impressive post-war rehabilitations in modern history, rivalled perhaps only by America's rebuilding of its own South after the US Civil War (1861–5). Germany and Japan, today, have massive free-market economies, and politically remain peaceful, stable, and decent democracies – good citizens on the global stage. And these countries are by no means "clones" (much less colonies) of America: each has gone its own way, pursuing political paths quite distinct from those that most interest the United States. Consider especially West Germany's journey, which went on to involve reunification with the former East Germany in 1990, and then the absolutely central role that the new Germany has carved out for itself within the European Union. So, *even though it will always be controversial to impose political values and new institutions*, there's evidence here that even massive, forcible post-war changes need *not* threaten "a nation's character." Yet it is fair to admit that for rehabilitation to work seems to require an enormous amount of time, money, and commitment.[33]

Rehabilitation More Recently, in Iraq and Afghanistan

It seems that America and its allies have, more recently, tried to implement many of the ten principles listed in table 6.1 in Iraq and Afghanistan. This was understandable, especially since these were drawn from the most successful cases: it's ideal to model ourselves after best practices, no? But the results, of course, have *not* been

Table 6.1 The Ten Principles of Rehabilitation

The occupying war winner, during post-war reconstruction, ought to:
  1. Adhere strictly to the laws of armed conflict during regime take-down and occupation.
  2. Purge most figures, symbols, and other items associated with the old regime, and prosecute its war criminals.
  3. Disarm and demilitarize the society.

But then
  4. Provide effective military and police security for the whole country.
  5. Work with a cross-section of locals on a new, rights-respecting constitution that features checks and balances.
  6. Allow other, non-state associations (i.e., civil society) to flourish.
  7. Forgo demanding compensation, and imposing sanctions, in favour of investing in and rebuilding the economy.
  8. If necessary, revamp educational curricula to purge earlier propaganda and cement new values.
  9. Ensure that the benefits of the new order will be (a) concrete, and (b) widely, not narrowly, distributed.
  10. Follow an orderly, not-too-hasty "exit strategy" once the new regime can stand on its own.

the same as they were for West Germany and Japan. Indeed, the more recent cases have seen a mixture of both (modest) successes and (substantial) failures. Let's consider these, and then explain the important differences between the two sets of cases regarding post-war rehabilitation.

Successes

In both countries, bad regimes implicated in aggression and terrorism were removed from power: the Taliban in Afghanistan (late November 2001); Saddam Hussein in Iraq (May 2003). New constitutions were written, and democratic elections were held several times to authorize new governments. Reparations and sanctions were not imposed. It's probably true that civil society, compared with what it was under Saddam Hussein or the Taliban, is freer and more active. The Kurds and Shi'a in Iraq, having been oppressed by Saddam, were no doubt thrilled to see him gone. Afghanistan has seen gains in gender equality, with the international community building and staffing new schools for girls and women, who went uneducated under the Taliban. International forces can also point to a range of concrete achievements, like the rebuilding of certain roads and securing such infrastructure as the oil supply in Iraq and food supplies into the cities of Afghanistan.[34]

The problem, though, is that the evidence suggests that it's *not* things like growth of civil society and gender and minority equality that matter most when it comes to the real-world success and durability of post-war reconstruction. Historical data suggest, rather, that the two most important things are: (a) *physical security* (i.e., personal safety from violence); and (b) *economic growth*. Jim Dobbins has distilled all the data into one crystal-clear rule of thumb regarding post-war success: the war-winning occupier and the new local regime have about 10–15 years (from "victory"/ end-of-main-war/start-of-occupation), to form an effective partnership and to make the average person in that society feel better off – more physically secure and more economically prosperous, especially – than they were prior to the outbreak of war. If they can do this, they'll probably succeed with post-war reconstruction; if not, there will be failure and a serious risk of backsliding into armed conflict.[35]

Using this rule of thumb, the outermost deadline for achieving this goal in Afghanistan was 2017, and in Iraq, 2018. The US-led occupation of both countries has been declared officially over, though – tellingly – many thousands of US troops today remain in both countries. Let's ask Dobbins's crucial question: in each country, how fully realized are both physical security and economic well-being for the average citizen?

Challenges: Security

Afghanistan continues to be rocked by political violence: even the capital Kabul has suffered hundreds dead from bombings in 2018 alone. The Afghan countryside is even less secure. Further, even though there have been elected governments, the reality is that they aren't stable. Their control, such as it is, doesn't extend much beyond Kabul, and there are many potent, rival factions: not just Taliban remnants but other terrorist groups. Indeed, in

March 2017, the US military – again, still there – made a big show
of dropping the MOAB ("Mother-Of-All-Bombs," the most
destructive non-nuclear bomb ever deployed) onto an Afghan
mountain range, with the goal of obliterating suspected terrorist
tunnels underneath. Also, as the US withdrew troops *en masse* from
Afghanistan, from 2012 to 2016, it very substantially stepped up
its drone attack program in that country. (More on drones in the
next chapter.) That makes good military sense – save your own
troops but still target enemy terrorists – but the issue remains: if
Afghanistan were truly safe and secure, then thousands of drone
strikes would presumably not be needed. So, *would the average
Afghan say that they feel more secure now than when the Taliban was in
power 17 years ago?* Perhaps not. This may be why US President
Donald Trump, in June 2017, authorized Defense Secretary James
Mattis to increase the number of US ground troops from 10,000
– a large number indeed "post-occupation," there to "advise and
train" the new Afghan military – to as many as 15,000. Would such
a surge of troops, if it happens, improve security in Afghanistan?
Or, rather, would it re-escalate a war that Dobbins tells us can no
longer be won? Time will tell.[36]

In Iraq, meanwhile, the security situation was so bad in 2005–6
that experts spoke openly of civil war between the three main
groups: the Kurds, Sunni, and Shi'a. At the time, US President
George W. Bush ordered 20,000(!) more US troops into Iraq, and
they succeeded in cutting down group-on-group violence. The
US then formally withdrew and officially ended its occupation in
2011. Since then, however, Iraq substantially fell apart and is today
a country perhaps in name only, with the Kurds pressing for full
autonomy in the north, with Sunni and Shi'a apparently never
to trust each other again, and with their antagonisms encouraged
further by rivalrous support from Saudi Arabia and Iran, respectively. Most tellingly, the radical Islamic terrorist group ISIS in
2014–15 took control of large chunks of the northern part of Iraq,
including some oil fields. Suffering major setbacks throughout
2017, ISIS's aim had been to install an extremely strict, Taliban-
style caliphate in that territory. The US still today has about 9,000
troops as "advisers" on the ground in Iraq, with most of their

focus on trying to help the new Iraq military, plus Kurdish militias, eliminate ISIS completely. So, would the average Iraqi say s/he is physically more secure than when Saddam was in power? Presumably all the Sunnis would answer no, whereas the Kurds and Shi'a might have more complex answers: they suffered under Saddam, and Saddam involved Iraq in very destructive wars with Iran (1980–89) and then the United States (1990–1 and 2003). But now that Iraq has all but fallen apart, and the Kurds, for instance, have been battling with ISIS over the north, it's quite unclear whether Iraq can be rebuilt in a thorough way, or whether whatever faction(s) eventually prevail will be any better than Saddam.[37]

Challenges: Economy
Would the average Afghan or Iraqi today feel more prosperous than prior to the war? Thankfully, the Americans didn't implement the retribution model in either case, and instead have sent some investment into both countries. Iraq is in the better position economically, as it has large reserves of oil and gas. Yet huge challenges remain. The near-constant war since 1979, plus the effects of the sanctions imposed from 1991 to 2003, devastated Iraq's basic infrastructure and well-being. So much rebuilding needs to be done, so many promising young people have been killed in the violence, and unemployment remains a terrible problem.[38]

Afghanistan, by contrast, is one of the world's poorest countries, with two-thirds of the population living on just two US dollars per day. The same proportion of the population is believed to be functionally illiterate, and unemployment is thought to afflict half the workforce. Afghanistan, like Iraq, faces issues of ruined infrastructure and the brutal consequences that constant warfare has inflicted on the economy. (These consequences can perhaps be condensed as follows: *would you open a business in a war zone?*) It remains unclear whether success will ever be achieved on this front. And this harsh fact remains: Afghanistan today has the second highest rate of emigration in the entire world, after Syria (with its dreadful civil war). If Afghanis truly felt like the future was going to get better, presumably they wouldn't be fleeing their country in search of a better future elsewhere. Here, too, it seems, the answer to the question, "Do you feel more prosperous?" is clear, and negative.[39]

6.3.4 Why the Differences in Rehabilitation Experiences?

Rehabilitation, as a post-war policy, has thus not worked nearly as well in Iraq and Afghanistan as it did in West Germany and Japan *because these are profoundly different situations.* It was incorrect to assume (as key players in the George W. Bush administration seem to have) that the same post-war policy could work equally well in very different circumstances. There are deep discordances between the two sets of cases:

- In the post–World War II cases, the *US had near-total control* over the occupied societies, as they had been so utterly destroyed.
- In the post–World War II cases, there was *no foreign interference* with US-led reconstruction, whereas in Afghanistan, and especially Iraq, such foreign powers as Iran have energetically tried to thwart US success.
- In the post–World War II cases, there was *no lingering resistance* to US occupation, whereas in both Iraq and Afghanistan there has been ferocious resistance, armed insurgency, and, in the case of Iraq, the next thing to civil war.
- In the post–World War II cases, there were *no substantial, bitter inter-group rivalries*: the people in West Germany and Japan were united in genuinely wanting US-led reconstruction to work. In Iraq and Afghanistan, by contrast, bitter rivalries and armed fighting between the local groups have made governance and peace-building nearly impossible.
- Finally, as has been well documented, *with WWII the US started post-war planning years in advance*, and the government knew what it wanted to do and how to do it in regard to Japan and West Germany. By contrast, in Afghanistan and especially Iraq, the post-war planning was hasty and haphazard, the local regimes fell much more quickly and easily than expected, and essentially the US tried rapidly to "default" to the WWII template instead of realizing how profoundly different the situations were.[40]

6.4 Eizenstat's "Gaps" Analysis

Speaking of resistance and insurgency, Stuart Eizenstat has crafted an influential "big picture" analysis of how to handle a violent insurgency, or generally how to bolster a weak or rebuilding state. He argues that violent insurgency arises from "gaps" which the current government – whether local, occupier, or a blend of both – has "let become open," or is having a hard time closing. Such gaps make the government look ineffective, and allow space for chaos and the continuation of conflict in spite of all the efforts to shut it down and turn the historical page.[41] These gaps are quite suggestive both of Dobbins's work, as well as the best practices rehabilitation "recipe." Three crucial elements determine whether there are such gaps or not: security, capacity, and legitimacy.

Eizenstat would define *security* as the ordinary person in that society feeling they're reliably safe from the threat of physical violence. They don't have to worry about such things on a regular basis, and can go about their daily lives. *Capacity* is Eizenstat's more minimal version of Dobbins's "economic growth/prosperity," referring more basically to a government being able to guarantee the vital material needs of people: supplies of water, food, electricity, housing, and core services like garbage collection and sewage treatment. And Eizenstat, and others like Charles Call, frame the above two requirements in terms of a social contract conception of justice. The average person asks themselves: Why should I obey a government if it can't provide me with basic physical security, or provide an environment in which my truly vital human needs can be met?[42] If the government can't do so, then rival players – like insurgent groups, militias, revolutionary groups – can aspire to step into that role and promise to do better, fueling a continuation of conflict. Finally, and here Eizenstat importantly goes beyond Dobbins and hooks into the best practices recipe (plus our thoughts at the end of chapter 4): *Legitimacy* refers to the ordinary person feeling that the government has the moral authority to govern, with sincere intentions to govern well on behalf of everyone, and not merely some "pet" insider group or tribe.[43]

Eizenstat argues that these three key elements are all profoundly interconnected. Violence (lack of security) can destroy infrastructure needed for capacity, for instance, and when a society lacks either, or both, of security and capacity, then everyone wonders about the government's legitimacy to govern, as its factual inability calls into serious question whether people should put their continued trust in it. Conversely, a government seen as illegitimate may elicit violent resistance to its rule, jeopardizing at least security and perhaps even capacity, depending on the scale of destruction. Crucially, the interconnections show that *severely attacking even just one of these elements will drag down everything, prolonging the conflict.* And, of course, the easiest and most tempting element to attack for insurgents, and opponents of peace deals generally, is physical security.[44]

It's tempting to draw an analogy to sports, based on the pith and suggestiveness of Eizenstat's analysis. The achievement of stable and decent post-war reconstruction may well be like "offence versus defense" in sports like football/soccer: all the defense has to do is disrupt and get in the way. It's a much easier, less careful, less patient job. Whereas, in such sports, often nearly everything has to come together in order for a goal to be scored. If compelling, it therefore follows that we should actually expect that *failure, or at least sub-optimality, are going to be the historical norm* when it comes to post-war settlements and attempted reconstructions.

6.5 Conclusion

Where does such a down-beat descriptive conclusion leave us, in terms of a plausible set of prescriptions for post-war justice? Let's draw the elements of this chapter together to suggest a reasonable conception of post-war justice, from a more revisionist JWT perspective, and incorporating items A through E:

[A] We should generally adhere to the principles of occupation law.

[B] Generally, we should not pursue retribution (or, at least, not pursue it as a general policy, whereas aspects of retribution called for within individual punishments, as mandated by war crimes trials, may be permissible).

[C] We must adhere to the five overlapping "thin" principles of post-war justice at its most basic: (1) ceasefire/armistice/ truce/cessation of hostilities; (2) exchange of POWs; (3) a public peace deal; (4) accountability (notably through well-structured war crimes trials); (5) proportionality.

[D] Generally rehabilitation seems desirable, and should be pursued. But when it comes to rehabilitation, while it's true that we need to have a healthy sense of what has actually worked in the past, we also need a healthy respect for the exact facts in the present situation, being mindful that the general "best historical practices recipe" might not always work, and/or needs to be carefully re-tailored to the case at hand.

[E] All best efforts at A–D should be made, with utmost focus on the crucial 10–15 years following the end of the war, guided especially by what we might call the three major, interlocking, institutional values of: security, capacity, and legitimacy.

It may be worth stressing how such a theory or conception of elements A through E isn't merely one targeted (as per realism) on a *stable and successful* post-war reconstruction, *but also on one that is just*, or aspires to satisfy moral and legal criteria. This is most straightforward when we draw directly on existing international law in A and C. More broadly, A through E can also be interpreted as trying to realize the human rights of people as best as can be done under the very difficult circumstances of post-conflict reconstruction. Recall, from the end of chapter 4, when discussing *jus ad bellum* and the right to govern, mention was made of five foundational human rights objects or claims: to life or personal security from violence; to individual freedom; to material subsistence; to equality, at least in the form of non-discrimination; and to recognition as a person and rights-bearer.[45] Our discussion in this chapter of rehabilitation, alongside Dobbins's research and Eizenstat's analysis, shows that such a rights-based conception of justice can be seen as residing behind, and informing and inspiring, the set of elements A through E. It may overlap with realism (though that's nothing to apologize for here, as you want it to actually work and endure), but it also aspires to offer a sincere vision of justice, which may win not only the support of JWT/LOAC but even serve as a reply to pacifism's challenge to offer something substantial, designed, amongst other things, to prevent further wars in the future. *Jus post bellum*, in other words, may have much in common not only with negative, but also positive, peace.[46]

7

The Future of Warfare

We've covered much ground in this book and turn now, at its conclusion, toward war's future. We've considered some of this already, for instance regarding how asymmetric forms of conflict are taking over from more classic, symmetrical forms. Or, more precisely, how current wars, like in Syria, tend to be *complex mixtures of both forms*, with a great many players involved, both foreign and domestic, each employing different tactics and strategy. Three major subjects remain for us here: new weapons and methods, cyberwar, and the hypothesis of a longer-term "solution" to warfare.

7.1 The More Things Change …

Reference to Syria makes us mindful of something salient. We're just about to discuss a bunch of brand new, cutting-edge, high-technology weapons and tactics, like drones, lasers, cyber, swarms, and automated, robotic systems. *As important as these are, it's critical not to lose sight of the fact that some things never change with war.* These would include: the tribalism; the struggle over power, governance, and legitimacy; and, in general, all the basic causes and drivers of war canvassed in chapter 1. But it's not just those "big picture" issues: it can be true for weapons and tactics, too. Consider that, today in Syria, the government there seems to be slowly winning its brutal civil war by engaging, amongst other things, in forms of siege warfare. This is one of the very oldest, cruelest, and most indiscriminate forms of armed conflict: surrounding a town/city/region entirely, blocking off all access in or out, and then either shelling/shooting/bombing everyone and everything inside, or else slowly starving it of vital resources until the people all surrender or die. Indeed, the Syrian conflict features such ancient and barbaric methods and technology *alongside* some of the latest and most advanced, including chemical gas and nerve agents, as well as tools of cyberwar. In many ways, Syria illustrates what acclaimed military historian John Keegan has long opined: if a war is long and costly enough, and the belligerents are enraged and committed enough, eventually, they will avail themselves of *any and all* tactics and technology – from primeval to space-age – which they believe will help them win. Realists, of course, would say: "We told you so." And the other theorists – whether pacifist or JWT/LOAC – will reply: "Perhaps, as a sad matter of fact but, as a matter of ideals and justice, look at the dreadful results of such a thing."[1]

7.2 Emerging Military Technologies and Tactics (EMTs)

Four *emerging military technologies* (EMTs) are especially prominent today. Cyberwar is one, which deserves its own separate section below, as it might not be merely one new tool *for* armed conflict but, further, signify *a new kind of conflict entirely*. First, let's deal with the others.

7.2.1 Non-lethal Weapons

These are otherwise known as *incapacitating agents*, which stun, shock, disable, or render unconscious but don't normally kill. Today's militaries are anxious to keep the body count low in war, as the public – especially in voting democracies – doesn't like to see large numbers of gruesome casualties. Incapacitating agents range from such everyday tools of law enforcement as pepper spray, tear gas, and Taser electric-shock guns to sophisticated, military-scale weaponry. The US military, for example, possesses a weapon ("The Dazzler") designed to cause permanent blindness to anyone within a certain distance of its deployment unless they are wearing specially calibrated and fitted eye-shields. It's an intensely bright laser discharge. The thinking is that enemies incapacitated in this way would promptly surrender, having their lives – though not their eyesight – saved. It's interesting and worth reflecting upon that, though the US has such a weapon, they've thus far chosen not to use it. There are also "concussion grenades" which mainly focus on delivering a very loud bang, and associated sound disorientation/incapacitation, but some of which also emit a light flash, albeit not one permanently blinding.[2]

7.2.2 Soldier Enhancements

Militaries know that they're constrained by the limits of the human body. This is especially true for the land-based army. Military officials are thus keenly interested in anything that will enhance soldier performance on patrol and especially in battle. Thus, technologies for providing enhanced vision, hearing, strength, and resilience are constantly being improved upon. Of special interest are drug-related enhancements designed to minimize the need for rest and sleep and to maximize battle-readiness, alertness, strength, endurance, pain-tolerance, and energy. The US military has developed (but doesn't use) a "souped-up" drug cocktail (containing, amongst other ingredients, forms of cocaine, adrenaline, caffeine, and testosterone) that allows some soldiers to maintain battle-ready energy, with no sleep, for 72 hours – three days! – straight. And it's not just short-term drug enhancements, or artificial prostheses, in view here: even research into genetic enhancement (i.e., deliberately manipulating one's genetic code to augment biological capabilities permanently) is on the table. There has long been interest, both theoretical and practical, in the possibility of a so-called super-soldier.[3]

7.2.3 "Unmanned" Weapons Systems: Drones, Swarms, and Killer Robots

"Unmanned systems" refer, in the main, to robots and drones. Militaries are increasingly using "autonomous," or semi-autonomous, weapons systems that are either controlled *directly* by robots/software or *operated from a distance by a human* via remote control using satellites and Global Positioning System (GPS) technology. The unmanned systems with the highest profile thus far are *drones*, which are very small planes that can often be flown remotely and without enemy radar detection, being used either for espionage and surveillance or else for dropping bombs and shooting missiles. The US military has rapidly escalated its use of drone technology over the past 20 years, especially for surveying rogue regimes (chapter 2) and attacking terrorist bases as part of the GWOT (chapters 1, 2, 4, and 5). To give a sense of the scale, and these numbers are under-reported if anything, since 2002, the US has deployed more than 600 drone strikes in Pakistan, Yemen, and Somalia, and well over 2,000 such strikes in Afghanistan. (Recall from the last chapter how drone attack escalation was the strategic accompaniment to the draw-down of human troops and "end-of-occupation" in Afghanistan. This is a graphic illustration, perhaps, of the very future of armed conflict: *a "swap-out" substitute, of drones for soldiers, of robots for people*.) Thousands of people have been killed in these drone strikes: mainly suspected terrorists but hundreds of unintended, collateral civilian casualties have been admitted (chapter 5, and more below).[4]

Drones have attracted controversy not only for that serious reason but also because, in the US, the president can simply order their use, and some say the temptations for secrecy, and the projection of war-like force without any kind of timely declaration (chapter 4), raises concerns about public accountability. (Indeed, the drone program was originally run in secret by the CIA and such concerns resulted in the transfer of control over to the US military.[5]) Supporters retort that drones: protect the lives of one's own soldiers; are very cost-effective (relative to the invasion of a manned force); convey technological sophistication and long-reach capability; and have had many successes, including pin-point assassinations of some of the world's most wanted terrorists.[6] There's an arms race in drones right now, across the world, owing to the comparative cheapness, mobility, aura of high-tech competence, and perceived targeting effectiveness of these weapons. The US Air Force has admitted that it's presently *training more drone pilots than any other kind of pilot*.[7] That's a further issue of controversy with this weapon: some say that, by creating enormous distance between the pilot and the battlefield, drones make killing "too easy." In addition, the use of drones may be at odds with what some, such as Michael Ignatieff, have called (chapter 5) the traditional "warrior ethos" of being willing to put oneself, as a soldier, at risk on the battlefield in defense of one's country. It can be jarring to note, in this regard, that the largest US drone pilot command center lies in the desert outside Las Vegas and has given rise to scathing comments about "commuter soldiers," "video gamers more than pilots," the unseemly location beside "Sin City," and so on. And, talk about distance between soldier and battlefield: between a bunker in Vegas and the border shared by Pakistan and Afghanistan. That's literally almost halfway around the world.[8]

Supporters of drone technology reply that this is simply where the technology is leading; and, what – are we supposed to *increase* the risks to *our own* soldiers? B.J. Strawser says that decent governments should care enough about their soldiers' safety to invest in technology which decreases such risks. And what does the "warrior ethos" actually demand? More precisely, realists say that such an ethos may be a romanticized mythology based on battles, and the state of military technology, of hundreds of years ago: *why should we still cling to that*, or preserve and even fetishize it, when the technology is becoming so utterly different? (See swarms, below!) As for "too much distance," and "making killing too easy," every weapon ever invented makes killing easier – so, how are drones especially culpable in this regard? Indeed, is there an ethical difference-in-kind between a drone and, say, an arrow? Plus, there might actually be something said *in favor of being detached from* the immediate situation, wherein fear, rage, chaos, pain, adrenalin, and/or confusion can cloud one's judgment, and lead to severe battlefield mistakes – whereas remote targeting systems can allow one to be more cool and dispassionate (assuming, admittedly, that they have been fed accurate information).[9]

One final exchange, between supporters and detractors of drones, and that has to do with their political impact. Critics argue that drones are weapons of escalation, and *ways to perpetuate conflict, especially in the GWOT,* as their targeting mistakes fuel outrage, and as people in the targeted countries resent the arrogance and technological superiority of those who can strike at will, from on-high. Daniel Brunstetter even describes drone warfare as "aerial occupation" of a country. Supporters reply that, again, superior weaponry has always elicited resentment from those lacking it, and so this isn't a new argument; plus, it's not clear that we should allow such emotional responses to block considered calculations of policy and best projection of force. What about critics who retort that such projection of force *encourages a sense of constant surveillance, of fear (or even terror) in the targeted communities* – of knowing that they might, at any time, be subject to a drone strike? Supporters, especially of drones as used to advance the GWOT, might get aggressive and counter-argue that *that is precisely the point*: to make suspected terrorists feel that they *are* being constantly surveilled, and *to fear suffering the very thing they seek to project elsewhere*; the sense that, at any time, they might be struck down with killing force. Drones, supporters conclude, might in that sense actually be seen as weapons not of escalation but of deterrence. The debate continues, of course, and raises many themes repeated with other kinds of novel, high-tech weaponry.[10]

One such kind concerns "swarming drones." The idea is, literally, of a dense, swirling swarm of hundreds or even thousands of tiny – hand-size – drone robots, swarming like a group of angry bees toward its target(s). The swarm robots may contain explosives, and/or just physically ram into their target(s), kamikaze-style. Moreover, the swarm robots would be run purely on software algorithms, a kind of artificial intelligence (AI), as controlling such an immense swarm in a coherent way would be impossible for any person, or even group of highly skilled people. The swarm, once launched, would be fully automated. The reason why it's getting so much attention and funding right now is because, if fully realized, *it'd become essentially impossible to defend against*: just way too many, way too small, moving way too fast, and too unpredictably, such that, even if you destroy a few, or even a few hundred, there would still be hundreds or even thousands left which would still hit the target(s) and be able to disrupt and destroy. Consider, by contrast, how nowadays most drones strike "from on-high," so that they won't be shot down by enemy forces on the ground. The swarm wouldn't need to concern itself with such things: raw, overwhelming quantity, so to speak, would translate into a high-quality strike. Imagine, as a thought experiment depicted in the documentary *Remote Control War*, such a swarm being unleashed by tech-savvy terrorists, or an enemy military, on a dense urban area, like Manhattan. The consequences could be catastrophic. And don't think the technology is too far afield. In January 2017, the US Navy successfully tested a swarm of 103 drone robots in the desert skies above California, and released a fascinating video of it – heavily censored – now available on YouTube. Beyond mere testing, a small swarm of 13 AI drones was unleashed *under real conditions* against a Russian airbase in Syria in early 2018, though apparently without success, and it's unclear who did the launching (though reasonable suspects may include Israel, America, and/or perhaps even Saudi Arabia in some capacity).[11]

The concern over semi-autonomous, or fully autonomous, weapons systems is growing. In addition to all the pros and cons mentioned above with the current generation of drones, there's further emphasis, especially with fully autonomous weapons like the swarms, about *the lack of human control and oversight over such things, once launched or put into the field*. The idea of a machine or software making decisions about killing humans is one that some people find very frightening – and, with earlier prototypes, there were cases of totally mistaken identification – while supporters suggest that (once fully developed) automated processes might actually be *more moderate and reasonable than humans*, emphatically during such pressured, emotionally super-charged situations like battle. In any event, there's been for the past few years an ongoing political campaign for a new international treaty "banning all killer robots." It had a high-profile series of meetings in April 2018 at UN offices in Geneva. It has the support of some state governments (especially those, it must be said, *lacking* such technology), and prominent NGOs like Human Rights Watch.[12] From a JWT/LOAC perspective, new weapons always raise questions about whether new laws and rules are needed for dealing with them, or whether they can be covered under the existing principles analyzed in chapter 5. The same issue arises with cyberwar and its related technologies.

7.3 Cyber-warfare

7.3.1 Initial Definitions

Cyberwar is proving crucial, its import ever-growing. It can be defined, broadly and initially, as the aggressive use of advanced computer technologies in ways intended to harm the fundamental interests and/or basic rights of a political actor (either state or non-state). And this definition, partially by way of distinguishing between genuine cyberwar in our sense – between political actors – versus mere cyber-crime (e.g., identity and credit card theft) and real but lower-level "cyber-harms" (e.g., the recent spate of hackers holding hospital e-records "hostage," until they get an extorted payment in untraceable cryptocurrency like bitcoin).[13]

The question arises: is cyberwar merely a tactic to be used *within* old-school, physical, "kinetic" warfare, or is it *a new kind of conflict altogether*? The answer emerging seems to be both, hence its growing import. It's straightforward to see how "cyber" can be used as a complement to kinetic war: both as a tool of *intelligence-gathering*, and as a tool of *disorientation to distract* prior to a kinetic strike. In 2009, for example, Israel apparently used such means to mess with Syrian air defense radar, manipulating it so that it would seem as though Israeli warplanes weren't penetrating Syria's airspace to perform a bombing run on a target. Apparently, the Syrians didn't realize what was truly going on until the Israelis had made their strike, turned around, and were just about to re-enter Israeli air space.[14] Again, a clear illustration of cyber as tool: more contested is whether cyber denotes a different kind of war altogether.

Many experts think so. The US Pentagon, for instance, describes cyber as "the fifth dimension" of warfare itself, after: land, water, air, and space. In 2017, US Congress allocated a whopping budget of US$6.7 billion to cyber to "respond to, and repair damage from, cyber-attacks." It has, accordingly, created a new military department, USCYBERCOM, to defend American security interests in the cyber-world.[15] And consider the scale: the British government, for instance, reports that its government computer systems are targeted 1,000 times *each month* (i.e., more than 30 times *per day*) in concerted cyber-attacks carried out by raiders ranging from petty criminals and amateur hackers to rogue regimes and malevolent militaries. North Korea, for example, was deemed to be behind the "WannaCry" cyber-attacks which temporarily froze, and damaged, part of the British public health care system in May 2017.[16]

Some cyber-attacks, such as WannaCry, have been severe, with more detailed below. Do these count as a genuine cyberwar? Some say yes, others say perhaps the best example is what we know now about Russia's concerted efforts in the cyber-world, over the past 5–10 years, to systematically cyber-attack, destabilize, and harm the interests of countries it deems enemies or rivals around the world, notably some of its former "partners" in the Soviet Union (USSR), as well as the European Union (EU) (especially Britain, France, and Germany) and the USA. We'll return to this hypothesis after some further helpful definitions.

Cyberwar/Cyber-tactics involves the following forms:

- *Espionage* (or "spying") involving the use of computer technology (especially the Internet) to gather information that a country tries to shield as a matter of national security. Chinese cyber-strikes reportedly focus on espionage, especially of the commercial variety. Many American high-tech firms, such as Google, Microsoft, and various weapons companies, have complained of Chinese cyber-espionage attacks, which have been used to access extremely sensitive, high-security information, including, especially, information on product design and patents. The companies have pressed the US government to respond, and in April 2018 the Trump administration did levy some economic tariffs, which might be considered payback sanctions of a kind in this regard.[17]

- *Disinformation*, via computer technology, which is to say the spread of deliberate untruths or falsehoods, usually via social media, or else a more targeted, confusing bundle of garbled information, in a manner that harms the security interests of the target. The Israeli befuddlement of Syria's radar system, as above, would count as an example of the latter, and we'll return below to allegations surrounding Russia's general spread of damaging disinformation in connection with the UK's Brexit vote, and the US presidential election, both in 2016.[18]

- *Sabotage*, involving the use of computer technology to destroy or impair the operation of various systems integral to the basic interests of a political community. Systems vulnerable to sabotage include electricity and power; water and fuel distribution; computerized parts of manufacturing facilities; transportation systems, notably air or rail; banking and the stock market; health care providers; and even the Internet itself, or at least the most frequently used websites, Internet service providers, and most fundamental operating systems.[19]

The countries most frequently mentioned in connection with cyberwar technology include the US, the UK, China, France, India, Israel, North Korea, Pakistan, and Russia. Non-state actors make use of it too, ranging from terrorist groups like ISIS spreading propaganda to sophisticated forms of information-gathering, sharing, and disruption by "hacktivist" organizations like Anonymous and WikiLeaks.[20]

7.3.2 *High-Profile Cases of Cyber-Strikes*

To get a deeper sense of the stakes, consider a few more examples of some of the highest-profile cyber cases.

In 1982, at the height of the Cold War, a Canadian oil and gas company believed they had a Soviet spy in their midst. They contacted American military officials. Together, the Canadians and Americans hatched a plan: they'd let the presumed spy steal what he was after – a computerized control system for regulating the flow of oil and gas (which the Russians wanted supposedly to modernize their pipeline system in Siberia). But American intelligence personnel programmed the computer system with "a logic bomb," designed to make the pipelines malfunction and eventually explode after the system was implemented. This is exactly what happened, causing some loss of life and a substantial setback for vital Soviet infrastructure. This is thought to be one of the earliest cyber-strikes ever, and furthermore an important example of how cyber-attacks can result not merely in cyber-harm *but also* physical, even lethal, kinetic harm.[21]

In 2007, Russia and one of its neighboring countries, Estonia, became embroiled in a dispute over the latter's relocation of a Soviet-era war memorial in Tallinn, Estonia's capital. Many Estonians regarded the monument – ostensibly a memorial to soldiers who died during WWII – as a symbol of Soviet oppression of Estonians following the war. When the Estonian government, facing angry protests about the monument, had the memorial moved, Russia responded with a crippling cyber-attack that targeted the websites of the Estonian government, national media outlets, and the country's largest banks. For nearly a week, these institutions couldn't conduct any business online. Most ATMs and credit/debit systems went offline as well. Perhaps most disturbingly, the attack only ended once Russia decided to release its grip: nearly everyone else was caught flat-footed. It may interest readers to know that, since then, Russia has deployed similar cyber-attacks against every single one of its former "partner" countries in the former USSR.[22]

Perhaps the most serious of such Russian attacks was the

"NotPetya" cyber-assault on the Ukraine, which shut down electric power to more than 250,000 people in that country, in 2015–16. It was reported in March 2018 that, presumably using the expertise gained in all such earlier attacks, the Russians actually cyber-accessed the control systems of numerous US electric power plants and power grids in 2016–17, but didn't proceed to sabotage them. The Pentagon did, however, release a stark, sobering report about Russian cyber capabilities in this regard, wondering openly about future intentions and, of course, calling for protective changes.[23]

In 2010, Iran was attacked by a computer virus (nicknamed "Stuxnet") commonly believed to have been a joint creation of the United States and Israel. A piece of malware (malicious software), this sophisticated computer virus was planted in a German-made cooling component of one of Iran's nuclear reactors. When it was activated, the virus disabled the coolers, threatening over-heating and even potential meltdown of the reactor. This forced the Iranians to shut it down completely for an unspecified time. The goal, reputedly, was to set back Iran's progress toward developing nuclear weapons.[24]

7.3.3 Lack of Law: Leading to Realist Threats/Schemes, and to Idealistic Manuals

One might wonder how all this quite aggressive, dangerous activity is allowed to occur. We know what realists would say: there's vital new technology, and political actors are figuring out how to use it to sharply serve their interests. But is there no international law in connection with cyber? *The answer right now is no* or, at least, not directly. "The Big Three" cyber-powers of America, China, and Russia apparently did get together in 2011 for "talks about talks" for potentially crafting a treaty regarding permissible methods of cyber-warfare, modeled after traditional LOAC treaties (chapter 5). But the talks fell apart amidst angry, bitter mutual accusations, and not a word has been spoken since about it between them. Thus, there's no direct law at all, much as with the "killer robots" issue mentioned above.[25] (Technological innovation typically out-paces legal regulation.) But, two big things to note. First, more realistically, every country has declared publicly that *it will view any "severe" cyber-strike on it as a just cause for going to war* (i.e., kinetic war) in response; and second, more idealistically, concerned by the cyber-lawlessness, a group of about 20 top international and military lawyers got together in 2012, very symbolically in the capital of Estonia, Tallinn, and crafted a model cyberwar treaty. Called "The Tallinn Manual," it has 95 principles(!) to govern cyber-conflict. (But, again, it's a model treaty, not a treaty itself, albeit one drafted and offered by world-class experts in the LOAC.) Many of these Tallinn principles share themes with, or are directly analogous to, the principles of *jus in bello*, as well as occupation law, described in the last two chapters.[26] Non-combatant immunity, for example, widely deemed violated by Russia in its various cyber-strikes. Proportionality, too – and people have wondered about Stuxnet in that connection, given the potential risk of meltdown to the Iranian nuclear reactor. And certainly: public accountability, as cyber, as with drones, raises many concerns about heavy, consequential conflict being run in secret, unaccountable ways. There's a robust and ever-growing JWT literature on cyberwar debating such issues, and forwarding various proposals, as it strives to fill the normative vacuum left by the lack of any kind of LOAC treaty.[27]

7.3.4 Integrating Cyber with Prior Definitions

It may be helpful to connect all these examples and concepts regarding cyberwar back to our chapter 1 discussion of war's ontology. Recall that, there, our definition of war was: an actual, intentional, and widespread armed conflict between groups of people, especially politically motivated groups of people (whether state or non-state actors). How does cyber conflict, thus far detailed, fit this definition of war?

It seems to do so quite well. We've detailed examples of both national governments, and non-state actors like ISIS, using such measures: hence, the politically motivated groups part. And we've specified cases of actual, and not merely latent, cyber-attacks. Likewise, they're most certainly deliberate, deployed with the intention to harm the target, or at least to benefit the attacker (at least in relative terms, or at the very least in terms of gaining information and/or leverage). The only issue of fit between our chapter 1 definition and cyber seems to revolve around "widespread" as well as "armed" conflict.

With both aspects, it's really an issue of how one interprets the cyber-world, as well as how much it has come to affect our lives. Believers in the reality of cyberwar say that, because of the sheer pervasiveness of computer technology throughout our lives and institutions — emphatically in the developed world — deliberate attacks on cyber-structure *do* satisfy the criteria of being *both* widespread and armed. It seems hard to deny the widespread nature, e.g., of the "WannaCry" attack, hitting nearly the entire UK public health system, or "NotPetya," cutting electricity to a quarter-million Ukrainians. Indeed, experts like Julian Richards have noted how cyber-strikes might potentially be able to affect many *more* people, much *more easily*, than even huge mobilizations and deployments of a country's military. Systematic cyber-sabotage of the most vital basic infrastructure may actually be one of the ultimate WMDs.[28] Indeed, such attacks, and the fear thereof, have sparked deep re-thinking about what should be digital and connected, and what perhaps needs now to be "analog" or "air-gapped," disconnected from the online world. (Note, though, that the Iranian nuclear reactor was certainly the latter, but this didn't prevent the "Stuxnet" attack.)

Cyber-conflict counts as "armed" in the sense of: *using a man-made implement which makes harming both worse and easier.* And not just harm in the cyber-world: the harm can translate into straightforward, physical, bodily harm. The "WannaCry" attacks, for example, led to cancelled surgeries and harmed British patients; and, most clearly, the Siberian logic bomb explosion actually killed people. (Think, too, of sad, similar interpersonal cases wherein cyber-shaming and cyber-bullying have pushed vulnerable victims to actual suicide.) And one shouldn't minimize the "mere" cyber-harms, either: think, for example, of the damage to Estonia's economy, in terms of lost sales, just from that one week of having its banking system frozen. Also, as George Lucas notes: it may actually be the case that some cyber-harms may be *worse, even far worse*, than physical harms. It's an interesting analogous thought experiment, for example, to ask yourself which you would choose, if you were forced: either having someone punch you in the face, breaking your nose and jaw, or else having someone steal your life savings, via cyber-theft from all your accounts? Perhaps the most telling thing in this regard – of the sheer seriousness and/or functional equivalence – is precisely the fact that every government in the world has publicly declared that any severe cyber-strike on it *will be considered the same as* a physical invasion and kinetic attack with troops, tanks, jets, and missiles. "The times, they are a'changing," as they say – and our concepts and definitions of warfare need to be flexible enough to keep up.[29]

7.3.5 Russia's Cyber Cold War?

Let's return to the hypothesis that Russia's activities over recent years may illustrate a committed, systematic cyberwar strategy (as opposed to various "one-off" strikes or attacks). Consider that Russian President Vladimir Putin has long said that "the worst event in modern history" was the collapse of the USSR, and its defeat in the Cold War (circa 1989–91) at the hands of the West, in particular: the US, the UK, France, Germany, and the EU collectively. Putin has been either president or prime minister of Russia since 1999 (19 years and counting!), and thus has had a long time indeed to implement an agenda. He's clearly Russia's "strongman" leader, and has a long-term plan to bolster Russia (back to unequivocal great power status, as per chapter 2) and to weaken the US and the EU, both as "payback" for the Cold War, as well as to augment, in relative terms, Russian expansion and power.[30]

Recall from the above that, from 2006 to 2014, Russia engaged in serious cyber-attacks on every one of its former "partner" countries in the USSR. Experts, with hindsight, now think Russia was "war-gaming" the technology: perfecting it, under real conditions (but against tiny countries who wouldn't dare to respond), so that Russia would then know how to use it better against the EU and the US, which it then started to do from 2014 to today. In the short term, perhaps this was a response to EU and US sanctions over

Russia's 2014 "annexation" of the Crimea, and its aggressive moves against the Ukraine, on behalf of the Russian minority there.[31]

Others believe there's a longer-term agenda. Sun Tzu, ancient author of the classic *The Art of War*, once proclaimed that it would be the ultimate art of war if one could figure out *how to turn one's opponent's major strengths into a weakness*. And one of the great strengths of the West, of course, is its technology, its openness, its globalized interconnectedness – exemplified by the Internet. So, this more pervasive perspective views the Russians as daringly asking: can we use all that to harm the West?[32] What you see today, according to this theory, are precisely the fruits of that strategy over the past 2–3 years, in at least 3 big ways:

- *Brexit.* We know now that Russia spread much *disinformation, via social media,* toward the "soft" or "swing" voters in lower/middle-class UK, promoting Brexit. While the causal impact of Russian cyber-intervention might be wondered at – certainly, other forces were at play, too – the actions themselves were real, the intent negative. There's little doubt that, at the least, Putin isn't displeased with Brexit, as it'll impose costs on, and distract the attention of, both the UK and the EU.[33]
- *Elsewhere in Europe.* We know that Russia has been: funding "far-right," strongly nationalist and anti–immigrant political parties; funding and organizing rallies for such groups; and spreading online related, sensationalized "fake news." And, while again we might query the degree of causal impact, and admit to other causes, it's hard to deny that the Kremlin can't be upset by the clear resurgence of far-right parties and policies within such countries as Austria, Italy, Hungary, and Poland. The result, after all, is a conflicted, even fractured EU – and many headaches for such champions of EU unity as German Chancellor Angela Merkel.[34]
- *United States.* There's now evidence, notably from the official Robert Mueller investigation, that Russia used its cyber-methods to try to influence the 2016 US presidential election. (Indeed, Mueller issued indictments for 12 Russian intelligence officers in July 2018, though they may never be handed over by Putin

to face trial.) Two forms of interference are alleged: espionage on Hillary Clinton's campaign (apparently with WikiLeaks' complicity); and disinformation in the form of crude, false, and manipulative anti-Clinton social media stories, tweets, and memes. Brad Allenby has referred to the latter as one kind of "weaponized narrative": made-up stories, fake news, designed with deliberate malign intent to sow hostility and political division in a target community.[35]

The background concept with this theory is that *it's a new kind of war, with a new kind of purpose.* Physically violent, kinetic war is very expensive and fraught with severe risks (witness Iraq and Afghanistan, and the GWOT more broadly). Engaging in such actions against well-endowed belligerents like the UK, France, and especially the US isn't such a wise idea. Especially not when you're a sitting government, a fixed target as it were (unlike the terrorists with their guerilla tactics, who can slip one's grasp). But cyberwar *will* harm their interests with: (1) no physical casualties to your side; (2) much lower costs than physical invasion; (3) much better chances of denial, misdirection, and obfuscation succeeding; and (4) advancing a new objective. This new objective is *not* of "defeating" your enemy but, over the long term, constantly harassing them, imposing costs on them, confusing them, driving up internal divisions amongst them – and that leaves you with *more space to yourself,* greater relative power and influence. Pure, yet highly sophisticated and utterly contemporary, realism from the Russians in this regard. This theory, such as it is, speculates the existence of an ongoing cyber Cold War and, obviously, Putin's Russia aspires to a different outcome this time. We'll see how the future unfolds.[36]

7.4 A Glimmer of Hope? The Democratic Peace Thesis

It's a resurgent time, nowadays, for realist nationalism. This has been shown in many ways throughout this book, and emphatically this chapter with Syrian sieges, arms races for drones, killer robots, super-soldiers, attacking swarms, as well as with cyberwar and great power "grand strategy" scheming. There may not be much to hope for right now but, here at book's end, let's try. Let's return to the idea mentioned in chapter 3, regarding pacifist hopes for an enduring, institutional solution to the problem of war.

Inspired by the Enlightenment, many historical thinkers crafted proposals for solving the problem of war altogether. After all, if critical rationality could create empirical science – as in Isaac Newton's great Enlightenment breakthrough, with the founding of modern physics in 1687 – and thereby dispel superstitious tradition, then perhaps reason could also solve the age-old problem of armed conflict. Many proposals for "ending war" were crafted, notably including those of: William Penn (1697); Abbé de Saint-Pierre (1712); and, indirectly, Jean-Jacques Rousseau (1750s). The great German philosopher Immanuel Kant follows in their train, no doubt inspired by Rousseau, of whom he thought very highly. Indeed, Kant's most famous essay on armed conflict is entitled *Perpetual Peace* (1795), and includes Kant's own proposals in this regard, widely agreed to be the most advanced and important of the so-called perpetual peace tradition.[37]

In his essay, Kant advanced a provocative thesis, namely, that "republican governments" (as he labeled them) *would not go to war against each other*. We should thus try to increase the number of these governments, as the very best way to increasing peace in the world. (He said that, ultimately, all such governments should join together in a voluntary "cosmopolitan federation" spanning the globe.) Kant's republican government is one familiar to us today. It respects and realizes the rights of its individual citizens, and governs with their consent. It's a limited government with no tyrannical designs, either against its own citizens' freedom or against the territory or authority of any foreign government. It has a free economy, urges its citizens to excel culturally and economically, and shows foreign visitors warm hospitality. Kant reasoned that a country like this would never start a war against another, similarly structured, country. It wouldn't be domineering and tyrannical, and thus not a conquering force. And its people, since they would have control, would never authorize such wars in the first place. Kant believed that people are fundamentally rational, so they don't want to start wars, which are so risky and destructive. Moreover, people living in a free society have all kinds of better ways to spend their time, to seek satisfaction, and to quench any competitive striving: arts, business, culture, education, the professions, personal romance, sports – you name it. These are the things that would occupy their time and thoughts instead of political conquest and territorial expansion.[38]

American political scientist Michael Doyle has, in our time, taken over Kant's early conjectures and advanced a "democratic peace thesis," suggesting that *democracies have never gone to war against each other, nor will they ever do so*, for reasons very similar to what Kant suggests. Note that Doyle's thesis, like Kant's, is only that republics/democracies won't go to war *against each other* – not that they won't go to war at all. In fact, Doyle notes that democracies can actually be quite belligerent when confronting non-democratic regime types, especially dictatorships, perhaps as they're convinced of their own moral superiority. Looking at the facts, Doyle has a point. Counting all major wars of the modern era (since 1750), the three countries most frequently involved in armed conflict have been the United States, the UK, and France. But, these countries, since becoming true democracies after World War I (once slavery had ended, and women had won the vote), have, indeed, never gone to war against each other. And the same goes for all the other democracies.[39]

Doyle's democratic peace, or "liberal peace," thesis has received considerable scrutiny and scholarly support. Many have tried to prove him wrong, offering up possible counter-examples. But these have been proven false (e.g., the War of 1812 between the UK and US, as neither were true democracies at that point) and there's now widespread consensus that the democratic peace thesis, or something like it, seems valid. If so, it may point to one promising and substantial way toward Kant's dream of "perpetual peace": if democracies never go to war against each other, then we need to increase the number of democracies world-wide so that we'll have more peace. Theoretically, *if every country were to become a democracy,*

we might have a true and enduring solution to the problem of war.[40] Such
thoughts raise important, intriguing speculations regarding how
certain kinds of regime are much more likely to lead to better lives
and futures, whereas other kinds of regime, and choices about
governance, make misery and dystopia a more likely outcome. To
put it at maximum pith, in terms of war: the liberal peace thesis is
rooted in a conviction that *the ultimate cause of armed conflict is bad
regimes* (and/or political players engaging in mass human rights
violations). It follows that the ultimate solution, if there is one,
would be to create good regimes – democratic, rights-respecting
ones – as far as the eye can see.

As nice, hopeful, and progressive as all that may sound, critical
questions might be raised, both factual and value-laden. On the
factual side, some realists have suggested that the true reason for
the democratic peace isn't the shared values and sincere mutual
support but, rather, *the hegemony of the United States amongst those
communities during the present era.* On the value-laden side, some
harken back to our chapter 1 discussion of the causes of war, and
see in the liberal peace proposal a version of what was there labeled
"the idealistic" cause for war: the clash over values, and the notion
that violence might somehow be used in service of some values
to create a better, more just world. This can easily be portrayed
as a naïve, even destructive belief: immediately conjuring up, for
example, thoughts of World War I as "the war to end all wars,"
or, indeed, "the war on behalf of democracy."[41] Rousseau, in his
relevant writings, warned way back in 1756 that the use of war
to establish a permanent peace (or a world government or federation deemed necessary to secure such) might actually be the most
menacing political belief of all, triggering a ton of bloodshed. He
said such a thing "may do more harm in a moment than it would
guard against for ages."[42]

Presumably, fans of the democratic peace thesis today don't
have to agree – certainly, Doyle himself doesn't talk aggressively
about going to war on behalf of democracy. Supporters of liberal
peace might, by contrast, assert that we're only entitled to pursue
such pro-democratic reforms in a small-scale and peaceful way,
resolving ourselves to a very long, and multi-faceted, "twilight

struggle" against the sort of corrupt regimes and nasty political actors that spark wars. Fair enough, yet the realists, here so to speak at the bitter end, would still note that the present moment doesn't offer up much hope for the success of such small-scale measures, and the slow spread of such values, given how sharply they are under fire: by tribal nationalism; by cyberwar; by extremists, terrorists, and strongmen of various sorts; and by scheming, rivalrous great powers. It all seems to strain against perpetual peace, and toward constant conflict.

Notes

Chapter 1

1 B. Orend, *The Morality of War* (Peterborough: Broadview, 2nd edn, 2013), 1–6.

2 D. Sorenson, *Syria in Ruins* (New York: Praeger, 2016).

3 J. Keegan, *The First World War* (New York: Vintage, 2000); J. Keegan, *The Second World War* (New York: Penguin, 2005).

4 H. Arendt, *On Violence* (London: Mariner, 2001).

5 T. Schachtman, *The Phony War, 1939–40* (London: Cotler, 1982).

6 M. Walker, *The Cold War: A History* (New York: Henry Holt, 1995).

7 A. Mumford, *Proxy Warfare* (Cambridge: Polity, 2013).

8 K. Holsti, *The State, War, and the State of War* (Cambridge: Cambridge University Press, 1996).

9 Orend, *Morality*, 1–6; war frequency statistics, and excellent interactive conflict maps, from: the Nobel Prize Organization, www.nobelprize.org; UCDP Conflict Encyclopedia, www.ucdp.se; and the Council on Foreign Relations: https://www.cfr.org/interactives/global-conflict-tracker#!/global-conflict-tracker.

10 C. von Clausewitz, *On War*, trans. A. Rapaport (Harmondsworth: Penguin Classics, 1995), 101–3. See also the Clausewitz comment

and selection at 127–31 of B. Orend, ed. of I. Kant's *On Perpetual Peace* (Peterborough: Broadview, 2015).

11 T. Abdullah, *A Short History of Iraq* (Toronto: Pearson, 2003); L. Freedman and E. Karsh, eds, *The Gulf Conflict, 1990–91* (Princeton, NJ: Princeton University Press, 1993); W. Murray and R. Scales, *The Iraq War* (Cambridge, MA: Harvard University Press, 2003).

12 B. Orend, *Introduction to International Studies* (Oxford: Oxford University Press, 2nd edn, 2018); T. Davies, *NGOs: A New History of Transnational Civil Society* (Oxford: Oxford University Press, 2014).

13 M. Axworthy, *A History of Iran: Empire of The Mind* (New York: Basic Books, 2nd edn, 2016).

14 D. Phillips, *The Kurdish Spring* (London: Routledge, 2015).

15 Jim Sterba, ed., *Terrorism and International Justice* (Oxford: Oxford University Press, 2003); J. Hughes, *Islamic Extremism and The War of Ideas* (Washington, DC: Hoover Institute, 2010); F. Gerges, *ISIS: A History* (Princeton, NJ: Princeton University Press, 2016).

16 P.W. Singer, *Corporate Warriors* (Ithaca, NY: Cornell University Press, 2008); J. Scahill, *Blackwater* (Washington, DC: Nation Books, 2008).

17 L. Kamienski, *Shooting Up: A Short History of Drugs and War* (Oxford: Oxford University Press, 2016); P. Watt and R. Zepeda, *Drug War Mexico* (London: Zed, 2012).

18 P. Charles, *Armed in America* (New York: Prometheus, 2018).

19 J. Dunnigan, *How to Make War* (New York: William Morrow, 4th edn, 2013); E. Simpson, *War from the Ground Up* (Oxford: Oxford University Press, 2016).

20 The details on weaponry in general, and division of military force, is indebted to: S. Biddle, *Military Power* (Princeton, NJ: Princeton University Press, 2006; C. Goltz, *The Conduct of War* (Carlisle, PA: The War College, 2015); and J. Keegan, *A History of Warfare* (New York: Vintage, 1994).

21 F. Allhoff, A. Henschke, and B.J. Strawser, eds, *Binary Bullets* (Oxford: Oxford University Press, 2016); M. Gunneflo, *Targeted Killing* (Cambridge: Cambridge University Press, 2016).

22 R.G. Grant, *Battle* (London: DK, 2005); US Central Intelligence Agency, www.cia.gov; Canadian Security Intelligence Service: www.csis-scrs.gc.ca.

23 Keegan, *A History of Warfare*, 2–31.

24 B.H. Liddell Hart, *Strategy* (London: Faber & Faber, 3rd edn, 1954).

25 R. Thornton, *Asymmetric Warfare* (London: Polity, 2007).

26 J. Joseph, *Hegemony: A Realist Analysis* (London: Routledge, 2007).

27 K. Waltz, *Realism and International Politics* (New York: Routledge, 2008).

28 Famed British philosopher and pacifist Bertrand Russell, for one, thought this was the only possible solution to the problem of war. See his *New Hopes for a Changing World* (London: Allen & Unwin, 1936).

29 S. Huntington, *The Clash of Civilizations and the Remaking of World Order* (New York: Simon & Schuster, 1996).

30 D. Welch, *Justice and the Genesis of War* (Cambridge: Cambridge University Press, 1988).

31 B. Orend, *Human Rights: Concept and Context* (Peterborough, ON: Broadview Press, 2002); Orend, *Introduction to International Studies*, 248–345.

32 E. Gellner, *Nations and Nationalisms* (Ithaca, NY: Cornell University Press, 2nd edn, 2009).

33 E. Saunders, "War and the Inner Circle," *Security Studies* (2015), 466–501; W. Danspeckgruber and C. Tripp, eds, *The Iraqi Aggression Against Kuwait* (Boulder, CO: Westview, 1996).

34 V. Lenin and N. Bukharin, *Imperialism and War* (London: Haymarket, 2010); F. Firet, *The Passing of an Illusion* (Cambridge, MA: Harvard University Press, 1999).

35 S. Pelletiere, *Iraq and the International Oil System: Why America Went to War in the Gulf* (London: Maisonneuve, 2nd edn, 2004); W.R. Clark, *Petrodollar Warfare: Oil, Iraq and the Future of the Dollar* (London: New Society, 2005).

36 J. Ledbetter, *Unwarranted Influence: Dwight Eisenhower and The Military–Industrial Complex* (New Haven, CT: Yale University Press, 2011); N. Turse, *The Complex: How the Military Invades Our Everyday Lives* (New York: Metropolitan, 2009).

37 Stockholm International Peace Research Institute (SIPRI): www.sipri.org.

38 S.M. Pavelec, ed., *The Military–Industrial Complex and American Society* (New York: ABC-CLIO, 2010).

39 SIPRI, see note 37; G. Crile, *Charlie Wilson's War* (London: Grove,

2007); S. Tanner, *Afghanistan: A Military History* (New York: De Capo, 2009).

40 J. Payne and A. Sahu, eds, *Defense Spending and Economic Growth* (Boulder, CO: Westview, 1993).

41 L. Keeley, *War Before Civilization* (Oxford: Oxford University Press, 1999); D. Peterson and R. Wrangham, *Demonic Males: Apes and the Origins of Human Violence* (New York: Mariner, 1997).

42 S. Faludi, *The Terror Dream* (New York: Metropolitan Books, 2007).

43 K. Alexander and M. Hawkesworth, eds, *War and Terror: Feminist Perspectives* (Chicago, IL: University of Chicago Press, 2008); J. Barrett, *History Matters: Patriarchy and the Challenge of Feminism* (Philadelphia, PA: University of Pennsylvania Press, 2007).

44 J. Meyer, *Men of War* (London: Palgrave Macmillan, 2012); T. Digby, *Love and War: How Militarism Shapes Sexuality and Romance* (New York: Columbia University Press, 2014).

45 Faludi, *The Terror Dream, passim*; C. Mackinnon, *Are Women Human?* (Cambridge, MA: Harvard University Press, 2007).

46 S. Freud, *Civilization and Its Discontents* (New York: Norton, 2005); B. Orend, *On War: A Dialogue* (New York: Rowman & Littlefield, 2009).

47 Keegan, *A History of Warfare, passim*; M. Gelven, *War and Existence* (Philadelphia, PA: Penn State University Press, 1994), 9–10.

Chapter 2

1 C. von Clausewitz, *On War*, trans. A. Rapaport (Harmondsworth: Penguin Classics, 1995); P. Morriss, *Power: A Philosophical Analysis* (Manchester: Manchester University Press, 2nd edn, 2002).

2 H. Morgenthau, *Politics Among Nations* (New York: Knopf, 5th edn, 1973).

3 T. Hobbes, *Leviathan*, ed. A.P. Martinich (Peterborough, ON: Broadview, 2004), Chapter 11, 79–80.

4 K. Waltz, *Man, The State and War* (Princeton, NJ: Princeton University Press, 1978), 4–5.

5 S. Forde, "Classical Realism" (62–84) and J. Donnelly, "Twentieth Century Realism" (85–111) both in T. Nardin and D. Mapel, eds,

Traditions in International Ethics (Cambridge: Cambridge University Press, 1992).

6 M. Walzer, *Just and Unjust Wars* (New York: Basic Books, 3rd edn, 2000), 3–15, with quote at 5.

7 G. Kennan, *Realities of American Foreign Policy* (Princeton, NJ: Princeton University Press, 1954); J.L. Gaddis, *Strategies of Containment* (Oxford: Oxford University Press, 2005).

8 R. Keohane, ed., *Neorealism and Its Critics* (New York: Columbia University Press, 1986).

9 R. Spegele, *Political Realism in International Relations* (Cambridge: Cambridge University Press, 2004); D. Boucher, *Political Theories of International Relations* (Oxford: Oxford University Press, 2001).

10 J. Nye, "Smart Power: In Search of the Balance between Hard and Soft Power," *Democracy* (Fall 2006), 21–46.

11 P. Kennedy, *The Rise and Fall of The Great Powers* (New York: Vintage, 1989).

12 N. Hyneck and D. Bosold, eds, *Canada's Foreign and Security Policy: Soft and Hard Strategies of a Middle Power* (Oxford: Oxford University Press, 2009); C. Holbraad, *Middle Powers in International Politics* (London: Macmillan, 1984).

13 K. Waltz, *Realism and International Politics* (New York: Routledge, 2008); B. Orend, *Introduction to International Studies* (Oxford: Oxford University Press, 2nd edn, 2018), Chapter 4.

14 H. Ebert and F. Flemes, eds, *Regional Powers and Contested Leadership* (London: Palgrave Macmillan, 2018); S. Mabon, *Saudi Arabia and Iran: Power and Rivalry in the Middle East* (London: I.B. Tauris, 2015).

15 J. Becker, *Rogue Regime* (Oxford: Oxford University Press, 2006).

16 Mabon, *Saudi Arabia and Iran, passim*; P. French, *North Korea: State of Paranoia* (London: Zed Books, 2nd edn, 2015); N. Chomsky, *Rogue States* (London: South End, 2000).

17 M. Bukovac, *Failed States: Unstable Countries in the 21st Century* (New York: The Rosen Group, 2011).

18 A. Ghani and C. Lockhardt, *Fixing Failed States* (Oxford: Oxford University Press, 2008).

19 N. Wheeler, *Saving Strangers* (Oxford: Oxford University Press, 2003); J. Holzgrefe and R. Keohane, eds, *Humanitarian Intervention* (Cambridge: Cambridge University Press, 2003).

20 J.C. Pevehouse and J. Goldstein, *International Relations* (New York: Pearson, 11th edn, 2016).

21 G. Berridge, *Diplomacy: Theory and Practice* (London: Palgrave Macmillan, 5th edn, 2015).

22 R. Haas and M. O'Sullivan, eds, *Honey and Vinegar: Incentives, Sanctions and Foreign Policy* (Washington, DC: Brookings Institution, 2000).

23 D. Drezner, *The Sanctions Paradox* (Cambridge: Cambridge University Press, 1999).

24 A.C. Drury, *Economic Sanctions and Presidential Decision-Making* (London: Palgrave Macmillan, 2nd edn, 2015); G.C. Hufbauer, J. Schott, K.A. Elliott, and B. Oegg, *Economic Sanctions Reconsidered* (London: PIIE, 3rd edn, 2009).

25 B. Early, *Busted Sanctions* (Palo Alto, CA: Stanford University Press, 2015).

26 G. Simons, *The Scourging of Iraq* (London: Macmillan, 3rd edn, 2014).

27 A. Pierce, "Just War Principles and Economic Sanctions," *Ethics and International Affairs* (1996), 99–113.

28 J. McMahan, "Realism, Morality and War" (78–92) and D. Mapel, "Realism and the Ethics of War and Peace" (180–200), both in T. Nardin, ed., *The Ethics of War and Peace: Religious and Secular Perspectives* (Princeton, NJ: Princeton University Press, 1996).

29 Morgenthau, *Politics, passim.*

30 M. Peterson, ed., *The Prisoner's Dilemma* (Cambridge: Cambridge University Press, 2015).

31 W. Morris, *News from Nowhere* (Peterborough, ON: Broadview Press, 2002).

32 H. Bull, *The Anarchical Society: A Study of Order in World Politics* (New York: Columbia University Press, 1977).

33 Kennan, *Realities of American Foreign Policy, passim*; S. Hoffmann, *Duties Beyond Borders* (Syracuse: Syracuse University Press, 1981).

34 T. Pogge, "The Bounds of Nationalism," in J. Couture, Kai Nielsen, and Michel Seymour, eds, *Rethinking Nationalism* (Calgary: University of Calgary Press, 1998), 463–504; C. Beitz, "Cosmopolitan Ideals and National Sentiment," *Journal of Philosophy* (1983), 591–600.

35 Donnelly, "Twentieth Century Realism," 85–111.

36 H. Kissinger, *Diplomacy* (New York: HarperCollins, 1995).

37 E.O. Wilson cited in Walzer, *Just and Unjust Wars*, 3–21; Clausewitz, *On War*, 3–6. Quote from W. T. Sherman, *From Atlanta to the Sea* (London: Folio Society, 1961), 121–2. See also E.A. Cohen, *The Big Stick: The Limits of Soft Power and the Necessity of Military Force* (New York: Basic Books, 2017).

38 Walzer, *Just and Unjust Wars*, 3–21.

39 Walzer, *Just and Unjust Wars*, 5–10, 15.

40 Walzer, *Just and Unjust Wars*, 251–68; B. Orend, *The Morality of War* (Peterborough, ON: Broadview Press, 2nd edn, 2013), 153–84.

41 J.D. Ohlin and L. May, *Necessity in International Law* (Oxford: Oxford University Press, 2016).

42 Cited in T. Nichols, *The Coming Age of Preventive War* (Philadelphia, PA: Penn State University Press, 2008), 2.

Chapter 3

1 M. Kaldor, *Human Security* (London: Polity, 2007).

2 S. Ruddick, *Maternal Thinking: Towards a Politics of Peace* (Boston, MA: Beacon, 2nd edn, 1995). This book bridges human security and pacifism with feminist analysis of war.

3 J. Teichman, *Pacifism and the Just War* (Oxford: Basil Blackwell, 1986), 3.

4 H. Arendt, *On Violence* (New York: Houghton Mifflin Harcourt, 1970).

5 A. Alvarez and R. Bachman, *Violence* (London: Sage, 2nd edn, 2013); R. Collins, *Violence* (Princeton, NJ: Princeton University Press, 2009).

6 P. Ackerman and J. DuVall, *A Force More Powerful* (New York: St. Martin's Press, 2000).

7 J. Narveson, "Pacifism: A Philosophical Analysis," in R. Wasserstrom, ed., *War and Morality* (Belmont, CA: Wadsworth, 1970), 63–78.

8 M. Ghandi, *Autobiography* (Boston, MA: Beacon, 1993).

9 J. Sterba, "Reconciling Pacifists and Just War Theorists," *Social Theory and Practice* (1992), 21–38.

10 L. May, *Contingent Pacifism: Revisiting Just War Theory* (Cambridge: Cambridge University Press, 2015).

11 M. Neu, "Why There is No Such Thing as Just-War Pacifism," *Social Theory and Practice* (2011), 44–68.

12 May, *Contingent Pacifism, passim*; the derivation judgment from A. Fiala in his excellent online entry on "Pacifism" at the *Stanford Encyclopedia of Philosophy*: https://plato.stanford.edu/entries/pacifism.

13 M.A. Fox, *Understanding Peace* (New York: Routledge, 2014).

14 Quote from Fiala at his "Pacifism," *Stanford Encyclopedia of Philosophy*. See also A. Fitz-Gibbon, *Positive Peace* (Amsterdam: Rodopi, 2010) and D. Cortright, *Peace: A History of Movements and Ideas* (Cambridge: Cambridge University Press, 1998).

15 A Fialia, *The Routledge Handbook of Pacifism and Nonviolence* (New York: Routledge, 2018).

16 T.N. Hanh, *Being Peace* (New York: Parallax, 3rd edn, 2005).

17 *Matthew* 6:44.

18 *Matthew* 11:29 and 26:52.

19 *Matthew* 5:9.

20 D. Dombrowski, *Christian Pacifism* (Philadelphia, PA: Temple University Press, 1991); L.S. Cahill, *Love Your Enemies: Discipleship, Pacifism and Just War Theory* (Minneapolis, MN: Fortress, 1994).

21 *Micah* 4:3.

22 B. Orend, *The Morality of War* (Peterborough, ON: Broadview Press, 2nd edn, 2013), 9–33.

23 The works of John Howard Yoder (e.g., *When War is Unjust: Being Honest in Just-War Thinking* (Minneapolis, MN: Augsburg, 1984)) offer a good example of a religious justification for pacifism. See also: M.C. Cartwright, "Conflicting Interpretations of Christian Pacifism," 197–213, and T.J. Koontz, "Christian Nonviolence: An Interpretation," 169–96, both in T. Nardin, ed., *The Ethics of War and Peace: Religious and Secular Perspectives* (Princeton, NJ: Princeton University Press, 1996). On JWT as part of official Catholic catechism, see <www.vatican.va/archive/catechism>.

24 R. Holmes, *On War and Morality* (Princeton, NJ: Princeton University Press, 1989); R. Holmes, *Pacifism: A Philosophy of Nonviolence* (London: Bloomsbury, 2017); R. Norman, *Ethics, Killing and War* (Cambridge: Cambridge University Press, 1995); A. Fiala, *Practical Pacifism* (New York: Algora, 2004); C. Ryan, "Self-Defense, Pacifism and Rights," *Ethics* (1983), 508–24.

25 G.E.M. Anscombe, "War and Murder," in R. Wasserstrom, ed., *War and Morality* (Belmont, CA: Wadsworth, 1970), 41–53.

26 A. Kasher, ed., *Ethics of War and Conflict* (London: Routledge, 2013).

27 P. Roots, *A Brief History of Pacifism from Jesus to Tolstoy* (Syracuse, NY: Syracuse University Press, 1993); P. Brock and N. Young, *Pacifism in the 20th Century* (Syracuse: Syracuse University Press, 1999).

28 Aristotle, *Nichomachean Ethics*, trans. W.D. Ross (Oxford: Oxford University Press, 1998); P. Foot, *Virtues and Vices* (Berkeley, CA: University of California Press, 1978).

29 S. Darwell, ed., *Virtue Ethics* (Oxford: Wiley-Blackwell, 2003).

30 M. Walzer, *Just and Unjust Wars* (New York: Basic Books, 3rd edn, 2000), xvi and 329–36.

31 W. James, "The Moral Equivalent of War," in R. Holmes, ed., *Nonviolence in Theory and Practice* (Chicago, IL: Waveland, 1990), 125–32.

32 Holmes, *On War and Morality*, 260–96; G. Sharp, *Waging Nonviolent Struggle* (Boston, MA: Porter Sargeant, 2005); P. Ackerman and J. DuVall, *A Force More Powerful* (New York: St. Martin's Press, 2000). There's a companion PBS documentary series by the same title.

33 G. Sharp, "The Technique of Nonviolent Action," in R. Holmes and B. Gan, eds, *Nonviolence in Theory and Practice* (Chicago, IL: Waveland, 2nd edn, 2005), 254.

34 J. Rawls, *A Theory of Justice* (Cambridge, MA: Harvard University Press, 1971), 370–82; Walzer, *Just and Unjust Wars*, 329–36.

35 W. Churchill and M. Ryan, *Pacifism as Pathology* (Winnipeg, MB: Arbeiter Ring Publishing, 1998).

36 J. Brown, *Gandhi and Civil Disobedience* (Cambridge: Cambridge University Press, 1977); D. Dalton, *Mahatma Gandhi: Nonviolent Power in Action* (New York: Columbia University Press, 1993); A. Fairclough, *To Redeem the Soul of America* (Athens, GA: University of Georgia Press, 1987); A. Morris, *The Origins of the Civil Rights Movement* (New York: Free Press, 1984); Ackerman and DuVall, *A Force More Powerful*, 61–113 and 305–35.

37 Walzer, *Just and Unjust Wars*, 330–5.

38 Ackerman and DuVall, *A Force More Powerful*, 207–41; L. Bergfelt, *Experiences of Civilian Resistance* (Uppsala: University of Uppsala Press, 1993); J. Thomas, *The Giant Killers* (New York: Taplinger, 1976).

39 Walzer, *Just and Unjust Wars*, xxii, 330–5.

40 Rawls is usually credited with the ideal/non-ideal theory distinction. See his *A Theory of Justice*, *passim*.

41 J. Bentham, *The Principles of Morals and Legislation* (Indianapolis, IN: Hackett, 1981); J.S. Mill, *Utilitarianism* (Indianapolis, IN: Hackett, 1987).

42 I. Bickerton, *Illusion of Victory: The True Costs of War* (Melbourne: Melbourne University Press, 2011).

43 H. Bull, *The Anarchical Society: A Study of Order in World Politics* (New York: Columbia University Press, 1977).

44 M. Walzer, "The Moral Standing of States: A Response to Four Critics," *Philosophy and Public Affairs* (1979/80), 209–29. Obama's eloquent speech, blending aspects of all our Big Three, at: https:// www.nobelprize.org/nobel_prizes/peace/laureates/2009/obama-lecture_en.html.

45 B. Orend, "Evaluating Pacifism," *Dialogue: Canadian Philosophical Review* (2001), 3–25.

46 J. Keegan, *The First World War* (New York: Vintage, 1992). Though, even there: Poland would, for instance, not be so critical of WWI, as its right to independence was reclaimed.

47 J. Keegan, *The Second World War* (New York: Vintage, 1990); M. Walzer, "World War II: Why was this War Different?," *Philosophy and Public Affairs* (1970/71), 3–21; G. Orwell, *1984* (Harmondsworth: Penguin, 1989).

48 Norman, *Ethics, Killing and War*, 80–93.

49 Narveson, "Pacifism," 62–77.

50 Holmes, *On War and Morality*, 146–213, esp. 183–213.

51 Anscombe, "War and Murder," 41–53.

52 Ryan, "Self-Defense," 508–24.

53 Holmes, *On War and Morality*, 146–213, esp. 183–213; Ryan, "Self-Defense," 508–24.

54 I. Kant, *On Perpetual Peace*, ed. B. Orend and trans. I. Johnston (Peterborough, ON: Broadview Press, 2015).

Chapter 4

1 A compact yet substantial sweeping history of just war theory from its origins to the present day can be found in Chapter 1 of B. Orend,

The Morality of War (Peterborough, ON: Broadview, 2nd edn, 2013), 9–33. The dean of historians of JWT is James Turner Johnson. See, e.g., his *The Just War Tradition and the Restraint of War* (Princeton, NJ: Princeton University Press, 1981). There are many edited volumes containing lots of original JWT articles, and which are historically organized from origins to today, including notably the mammoth collection: G. Reichberg, H. Syse, and E. Begby, eds, *The Ethics of War: Classic and Contemporary Readings* (London: Wiley-Blackwell, 2006).

For the history of the LOAC, see: A. Gillespie, *History of the Laws of War* (London: Hart, 3rd edn, 2011) and S. Neff, *War and The Law of Nations* (Cambridge: Cambridge University Press, 2008). On the content of the LOAC as they presently exist, see for excellent overviews: A. Roberts and R. Guelff, eds, *Documentation on the Laws of War* (Oxford: Oxford University Press, 3rd edn, 2000); G. Solis, *The Law of Armed Conflict* (Cambridge: Cambridge University Press, 2010).

2 P. Christopher, *The Ethics of War and Peace* (Englewood Cliffs, NJ: Prentice-Hall, 1994); T. Nardin, *The Ethics of War and Peace: Religious and Secular Perspectives* (Princeton, NJ: Princeton University Press, 1998).

3 M. Reisman and C. Antoniou, eds, *The Laws of War* (New York: Vintage, 1994); M. Walzer, *Just and Unjust Wars* (New York: Basic Books, 2nd edn, 1990).

4 V. Lowe, A. Roberts, J. Welsh, and D. Zaum, eds, *The United Nations Security Council and War* (Oxford: Oxford University Press, 2010).

5 John Keegan, *A History of Warfare* (New York: Random House, 1993); R.G. Grant, *Battle* (New York: DK, 2005).

6 Christopher, *The Ethics of War and Peace*, 10–11, citing Aristotle's *Politics* [1256b, 25] for the locution of a kind of warfare being just.

7 B. Orend, *War and International Justice: A Kantian Perspective* (Waterloo, ON: Wilfrid Laurier University Press, 2000).

8 B. Orend, *Human Rights: Concept and Context* (Peterborough, ON: Broadview, 2002).

9 J. Locke, *Two Treatises of Civil Government* (Cambridge: Cambridge University Press, 1988); A. Cassese, *International Law* (Oxford: Oxford University Press, 2nd edn, 2005).

10 Orend, *The Morality of War*, 33–70; H. Shue, *Basic Rights* (Princeton, NJ: Princeton University Press, 1980).

11 J. Narveson, "Pacifism: A Philosophical Analysis," in R. Wasserstrom, ed., *Morality and War* (Belmont, CA: Wadsworth, 1970), 63–77.

12 T. Hurka, "Proportionality and Necessity," in L. May, ed., *War: Essays in Political Philosophy* (Cambridge: Cambridge University Press, 2008), 127–44.

13 Walzer, *Just and Unjust Wars*, 49, xiv.

14 James T. Johnson and G. Weigel, eds, *Just War and the Gulf War* (Washington, DC: University Press of America, 1991).

15 Orend, *The Morality of War*, 10–11.

16 L. Fisher, *Presidential War Power* (Lawrence, KS: University of Kansas Press, 3rd edn, 2013).

17 Fisher, *Presidential War Power, passim.*

18 For detailed sourcing of treaty-based references to each of those four *jus ad bellum* rules, see Orend, *The Morality of War*, Chapter 2.

19 Orend, *The Morality of War*, 12–13; H.A. Deane, *The Social and Political Ideas of St. Augustine* (New York: Columbia University Press, 1963).

20 Orend, *The Morality of War*, 48–52.

21 Ibid., 60–1.

22 F. Miller and L. Varley, *300* (Milwaukie, OR: Dark Horse Books, 2004).

23 I. Crouch, *A Pyrrhic Victory* (New York: Strategic Books, 2015).

24 Der Spiegel Magazine, *Inside 9/11* (New York: St. Martin's Press, 2002).

25 J. Corbin, *Al-Qaeda* (New York: Nation Books, 2002); A. Rashid, *Taliban* (New Haven, CT: Yale University Press, 2001); J.B. Eshtain, *Just War Against Terror* (New York: Basic Books, 2003); R. Falk, *The Great Terror War* (New York: Olive Branch Press, 2002).

26 P. Berman, *Terror and Liberalism* (New York: W.W. Norton, 2003); M. Scheuer, *Osama Bin Laden* (Oxford: Oxford University Press, 2011); D. Filkins, *The Forever War* (New York: Vintage, 2009).

27 B. Woodward, *Plan of Attack* (New York: Random House, 2004); P. Shiner and A. Williams, eds, *The Iraq War and International Law* (New York: Hart, 2008).

28 Orend, *The Morality of War*, 78–82.

29 J. McMahan, "Preventive War and The Killing of The Innocent," in D. Rodin and R. Sorabji, eds, *The Ethics of War* (London: Ashgate, 2006), 169–90.

30 Orend, *The Morality of War*, 78–82.

31 Ibid., 71–110.

32 Walzer, *Just and Unjust Wars*, 74–85.

33 S. Ritter, *War on Iraq* (New York: Profile, 2002).

34 *The National Security Strategy of the United States of America*. September 2002, available at www.state.gov/documents/organization/63562. pdf, p. 15.

35 M. Isikoff and D. Corn, *Hubris* (New York: Random House, 2006); W. Murray and R. Scales, Jr., *The Iraq War* (Cambridge, MA: Harvard University Press, 2003); *The Congressional Commission Report on the Attacks of 9/11* (Washington, DC: US Congress, 2004).

36 A. Bacevich, *America's War for the Greater Middle East* (New York: Random House, 2017); F. Gerges, *ISIS: A History* (Princeton, NJ: Princeton University Press, 2016); J. Keegan, *The Iraq War* (New York: Knopf, 2004).

37 But, admittedly, not in every context, especially historically: there may well be civil wars where there isn't disproportionate, one-sided slaughter or grievous crimes like ethnic cleansing but, rather, a genuine bitter contest over sovereignty between local groups wherein foreign intervention may not be appropriate at all. Thanks to one of my readers, and generally for such more traditional cases, see Walzer, *Just and Unjust Wars*, 86–108.

38 G. Prunier, *Rwanda: History of a Genocide* (New York: Columbia University Press, 1995); M. Barnett, *Eyewitness to a Genocide: The UN and Rwanda* (Ithaca, NY: Cornell University Press, 2003); R. Dallaire, *Shake Hands with the Devil* (New York: Random House, 2004).

39 Orend, *The Morality of War*, 94–9.

40 N. Wheeler, *Saving Strangers* (Oxford: Oxford University Press, 2003); J. Holzgrefe and R. Keohane, eds, *Humanitarian Intervention* (Cambridge: Cambridge University Press, 2003).

41 Orend, *The Morality of War*, 94–9; M. Finnemore, *The Purpose of Intervention* (Ithaca, NY: Cornell University Press, 2003); T. Weiss, *Humanitarian Intervention* (London: Polity, 2nd edn, 2012).

42 The International Commission on Intervention and State Sovereignty,

Report: The Responsibility to Protect (Ottawa, ON: International Development Research Centre, 2002).

43 A. Bellamy, *Responsibility to Protect* (London: Polity, 2009); G. Evans, *The Responsibility to Protect: Ending Mass Atrocity Crimes Once and For All* (Washington, DC: Brookings Institution, 2009).

44 And UNSC resolutions are considered binding international law. N. Gvosdev, *R2P: Sovereignty and Intervention After Libya* (London: World Politics Review, 2011).

45 A. Bayat, *Revolution Without Revolutionaries: Making Sense of the Arab Spring* (Palo Alto, CA: Stanford University Press, 2017); L. Noueihad and A. Warren, *The Battle for the Arab Spring* (New Haven, CT: Yale University Press, 2012).

46 C. Chivvis, *Toppling Qaffadi: Libya and the Limits of Liberal Intervention* (Cambridge: Cambridge University Press, 2014).

47 Council on Foreign Relations, *Syria: The Crisis and Its Implications* (Washington, DC: United States Senate, 2012); R. Erlich, *Inside Syria* (London: Prometheus, 2016); D. Sorenson, *Syria in Ruins* (New York: Praeger, 2016).

48 Orend, *The Morality of War*, 33–70.

49 Ibid. Compare and contrast with Walzer, *Just and Unjust Wars*; J. Rawls, *The Law of Peoples* (Cambridge, MA: Harvard University Press, 1999) and B. Roth, *Governmental Illegitimacy in International Law* (Oxford: Oxford University Press, 2001).

50 Orend, *Human Rights, passim.*

Chapter 5

1 M. Walzer, *Just and Unjust Wars* (New York: Basic Books, 3rd edn, 2000), 40–156.

2 W. Reisman and C. Antoniou, eds, *The Laws of War: A Comprehensive Collection of Primary Documents on International Laws Governing Armed Conflict* (New York: Vintage, 1994); M. Howard, *The Laws of War: Constraints on Warfare in The Western World* (New Haven, CT: Yale University Press, 1994); S. Neff, *War and the Law of Nations* (Cambridge: Cambridge University Press, 2008).

3 H. Morgenthau, *Politics Among Nations* (New York: Knopf, 5th edn, 1973).

4 M. Ignatieff, *The Warrior's Honour* (New York: Henry Holt, 1998); S. French, *The Code of the Warrior* (New York: Rowman & Littlefield, 2004).

5 B. Orend, *The Morality of War* (Peterborough, ON: Broadview Press, 2nd edn, 2013), 111–52; A. Roberts and R. Guelff, eds, *Documents on the Laws of War* (Oxford: Oxford University Press, 3rd edn, 2000); A. Gillespie, *History of the Laws of War* (New York: Hart, 3rd edn, 2011).

6 Walzer, *Just and Unjust Wars*, 40–156; Orend, *The Morality of War*, 111–52; Reisman and Antoniou, eds, *The Laws of War*, 73–94; G. Solis, *The Law of Armed Conflict* (Cambridge: Cambridge University Press, 2010).

7 Walzer, *Just and Unjust Wars*, 40–156; Orend, *The Morality of War*, 111–52; Reisman and Antoniou, eds, *The Laws of War*, 73–94; G. Simpson, *Law, War and Crime* (Cambridge: Polity, 2007).

8 G. Herring, *America's Longest War* (New York: McGraw-Hill, 4th edn, 2001).

9 J. Olsen and R. Roberts, eds, *My Lai: A Brief History with Documents* (London: Palgrave Macmillan, 1998); P. French, ed., *Individual and Collective Responsibility: The Massacre at My Lai* (Cambridge, MA: Harvard University Press, 1972). Thanks to one of my readers.

10 Reisman and Antoniou, eds, *The Laws of War*, 293–316; J.B. Elshtain, *Just War Against Terror* (New York: Basic Books, 2003).

11 M. Scheuer, *Osama Bin Laden* (Oxford: Oxford University Press, 2011); P.J. Whittaker, ed., *The Terrorism Reader* (London: Routledge, 2012).

12 T. Nagel, "War and Massacre," *Philosophy and Public Affairs* (1971/72), 123.

13 Orend, *The Morality of War*, 71–8 and 111–52; J. Sterba, ed., *Terrorism and International Justice* (Oxford: Oxford University Press, 2003).

14 W. Clark, *Waging Modern War* (New York: Public Affairs, 2002); M. Ignatieff, *Virtual War* (New York: Metropolitan, 2000); W. Murray and R. Scales, *The Iraq War* (Cambridge, MA: Harvard University Press, 2003).

15 J. Keegan, *A History of Warfare* (New York: Vintage, 1994), 13–18; C. Goltz, *The Conduct of War* (Carlisle, PA: The War College, 2015).

16 F. Rosen, *Collateral Damage* (Oxford: Oxford University Press, 2016).

17 I. Primoratz, ed., *Civilian Immunity in Wartime* (Oxford: Oxford

University Press, 2010); D. Kennedy, *Of War and Law* (Princeton, NJ: Princeton University Press, 2006).

18 R. Holmes, *On War and Morality* (Princeton, NJ: Princeton University Press, 1989).

19 P.A. Woodward, ed., *The Doctrine of Double Effect* (Notre Dame, IN: University of Notre Dame Press, 2001).

20 F.M. Kamm, *The Moral Target* (Oxford: Oxford University Press, 2012).

21 Walzer, *Just and Unjust Wars*, 106.

22 Orend, *The Morality of War*, 121–4.

23 J. Rawls, *The Law of Peoples* (Cambridge, MA: Harvard University Press, 1999), 98–105; F.X. Winters, *Remembering Hiroshima: Was It Just?* (New York: Routledge, 2016).

24 Orend, *The Morality of War*, 111–52; T. Hurka, "Proportionality in the Morality of War," *Philosophy and Public Affairs* (2005), 34–66; J. McMahan, *Killing in War* (Oxford: Oxford University Press, 2nd edn, 2011); D. Rodin, *War and Self-Defence* (Oxford: Oxford University Press, 2005).

25 J. Inciardi and L. Harrison, eds, *Harm Reduction: National and International Perspectives* (London: Sage, 1999).

26 Orend, *The Morality of War*, 273–96; Walzer, *Just and Unjust Wars*, 330–5.

27 P. Singer, *Children at War* (Berkeley, CA: University of California Press, 2006); M. Green, *The Wizard of the Nile* (London: Portobello, 2008). See also E. Amony, *I am Evelyn Amony: Reclaiming My Life from the LRA* (Madison, WI: University of Wisconsin Press, 2015) as well as the documentary on female child soldiers, "*Grace, Milly, Lucy.*"

28 R. Dallaire, *They Fight Like Soldiers, They Die Like Children* (Toronto: Vintage, 2010); Amnesty International: http://www.amnesty.org/en/news/landmark-icc-verdict-over-use-child-soldiers-2012-03-14; U. Jha, *Child Soldiers: Practice, Law and Remedies* (New Delhi: Vij, 2018).

29 A. Krammer, *Prisoners of War* (New York: Praeger, 2007); G. Best, *War and Law Since 1945* (Oxford: Clarendon, 1994).

30 Reisman and Antoniou, eds, *The Laws of War*, 41–7, 149–230, M. Byers, *War Law* (Washington, DC: Atlantic, 2009); F. Borch, *Geneva Conventions* (New York: Kaplan, 2010).

31 E. Saar, *Inside the Wire* (New York: Penguin, 2005); S. Hersh, *Chain of Command: The Road from 9/11 to Abu-Ghraib* (New York: HarperCollins, 2004); US Senate Select Committee on Intelligence, *Report on the CIA's Detention and Interrogation Program*: https://www. intelligence.senate.gov/study2014/sscistudy1.pdf.

32 D. Jinks, *The Rules of War* (Oxford: Oxford University Press, 2013). M. Danner, *Torture and Truth: America, Abu-Ghraib and the War on Terror* (New York: New York Review of Books, 2004). B. Innes, *The History of Torture* (New York: St Martin's Press, 1998); Y. Ginbar, *Why Not Torture Terrorists?* (Oxford: Oxford University Press, 2010).

33 Walzer, *Just and Unjust Wars*, 129; T. Hurka, "Proportionality and Necessity," in L. May, ed., *War: Essays in Political Philosophy* (Cambridge: Cambridge University Press, 2008), 127–44.

34 Reisman and Antoniou, eds, *The Laws of War*, 35–132; W. Boothby, *Weapons and the Laws of Armed Conflict* (Oxford: Oxford University Press, 2009). Thanks, re "taboo," to Prof. Dr. Milos Vec of the University of Vienna.

35 J. Cirincione, J.B. Wolfsthal, and M. Rajkumar, *Deadly Arsenals* (New York: Carnegie Endowment for International Peace, 2nd edn, 2005); N. Busch, *Combating WMDs: The Future of Non-proliferation* (Atlanta, GA: University of Georgia Press, 2009).

36 Boothby, *Weapons, passim*; J. Siracusa and A. Warren, *Weapons of Mass Destruction: The Search for Global Security* (New York: Rowman & Littlefield, 2015).

37 B. Allen, *Rape Warfare* (Minneapolis, MN: University of Minnesota Press, 1996); A. Stigylmayer, ed., *Mass Rape* (Omaha, NE: University of Nebraska Press, 1994); J. Leathermore, *Sexual Violence and Armed Conflict* (Cambridge: Polity, 2011).

38 D.K. Cohen, *Rape During Civil War* (Ithaca, NY: Cornell University Press, 2016); K. Crawford, *Wartime Sexual Violence* (Washington, DC: Georgetown University Press, 2017).

39 Ignatieff, *The Warrior's Honour, passim*; French, *The Code of the Warrior, passim*.

40 Y. Dinstein, *The Conduct of Hostilities under the Law of International Armed Conflict* (Cambridge: Cambridge University Press, 2010).

41 Walzer, *Just and Unjust Wars*, 207–22; J. Gardner, *The Blitz* (London: General, 2011); F. Taylor, *Dresden* (New York: HarperCollins, 2004).

42 Walzer, *Just and Unjust Wars*, 251–68.

43 B. Orend, "Is There a Supreme Emergency Exemption?" in M. Evans, ed. *Just War Theory: A Reappraisal* (Edinburgh: Edinburgh University Press, 2005), 134–56.

44 F. Chalk and K. Jonassohn, eds, *The History and Sociology of Genocide* (New Haven, CT: Yale University Press, 1990); S. Totten, W.S. Parsons, and I.W. Charny, eds, *Century of Genocide* (New York: Garland, 1997).

45 A. Bellamy, *Responsibility to Protect* (London: Polity, 2009); R. Dallaire, *Shake Hands with the Devil: Humanity's Failure in Rwanda* (New York: Random House, 2003).

46 Walzer, *Just and Unjust Wars*, 251–68.

47 K. Kessler Ferzan and L. Alexander, *Crime and Culpability* (Cambridge: Cambridge University Press, 2009).

48 Orend, *The Morality of War*, 153–71.

49 D. Rodin and H. Shue, *Just and Unjust Warriors* (Oxford: Oxford University Press, 2008); McMahan, *Killing in War*, *passim*; Rodin, *War and Self-Defence*, *passim*.

50 Walzer, *Just and Unjust Wars*, 38–40, 127–38; Reisman and Antoniou, eds, *The Laws of War*, 35–57; Y. Benbaji, "A Defense of the Traditional War Convention," *Ethics* (2008), 464–95; U. Steinhoff, *On the Ethics of War and Terrorism* (Oxford: Oxford University Press, 2007).

51 Rodin and Shue, *Just and Unjust Warriors*, *passim*; A.A. Haque, *Law and Morality at War* (Oxford: Oxford University Press, 2017); H. Frowe, *The Ethics of War and Peace: An Introduction* (London: Routledge, 2nd edn, 2015); S. Lazar and H. Frowe, eds, *The Oxford Handbook of Ethics of War* (Oxford: Oxford University Press, 2018).

52 Orend, *The Morality of War*, 113–16.

Chapter 6

1 B. Orend, *War and International Justice: A Kantian Perspective* (Waterloo, ON: Wilfrid Laurier University Press, 2000), 217–66.

2 J.T. Johnson, *The Just War Tradition and the Restraint of War* (Princeton, NJ: Princeton University Press, 1981); Seth Lazar's "War" entry in the *Stanford Encyclopedia of Philosophy*: https://plato.stanford.edu/entries/war/.

3 W. Reisman and C. Antoniou, eds, *The Laws of War* (New York: Vintage, 1994), 130–3.

4 E. Benvinisti, *The International Law of Occupation* (Oxford: Oxford University Press, 2nd edn, 2012); E. Carlton, *Occupation* (New York: Routledge, 1992); Y. Dinstein, *The International Law of Belligerent Occupation* (Cambridge: Cambridge University Press, 2009).

5 Benvinisti, *Occupation, passim*; Dinstein, *Belligerent Occupation, passim.*

6 For more on how violence can continue even after war ends, see M. Boyle's *Violence After War: Explaining Instability in Post-Conflict Societies* (Baltimore, MD: Johns Hopkins University Press, 2014).

7 M. Walzer, *Just and Unjust Wars* (New York: Basic Books, 3rd edn, 2000), 288.

8 J. Goldsmith and E. Posner, *The Limits of International Law* (Oxford: Oxford University Press, 2005); G. Simpson, *Law, War and Crime* (Oxford: Polity, 2007).

9 Walzer, *Just and Unjust Wars*, 292–301.

10 S. Neff, *War and the Law of Nations* (Cambridge: Cambridge University Press, 2008); D. Rodin and H. Shue, *Just and Unjust Warriors* (Oxford: Oxford University Press, 2008).

11 A. Roberts and R. Guelff, eds, *Documents on the Laws of War* (New York: Oxford University Press, 3rd edn, 2000); A. Gillespie, *History of the Laws of War* (New York: Hart, 3rd edn, 2011).

12 T. Tanaka, *Hidden Horrors: Japanese War Crimes in World War II* (Boulder, CO: Westview, 1997); T.P. Maya, *Judgment at Tokyo: The Japanese War Crimes Trial* (Lexington, KY: University of Kentucky Press, 2001).

13 G. Hankel, *The Leipzig Trials* (Dordrecht: Republic, 2014). Thanks to Prof. Dr. Milos Vec of the University of Vienna.

14 G. Robertson, *Crimes Against Humanity* (London: Penguin, 2012).

15 Admittedly, however, there was a nearly 100% conviction rate in the Tokyo trials, for hundreds of accused, which perhaps rightly raises eyebrows. See Maya, *Judgment at Tokyo, passim*. For Nuremberg: J. Persico, *Nuremberg: Infamy on Trial* (New York: Penguin, 1995); M. Marus, *The Nuremberg War Crimes Trials 1945–6* (New York: St. Martin's Press, 1997).

16 Reisman and Antoniou, eds, *The Laws of War*, 318–37; Persico,

Nuremberg: Infamy on Trial, passim; Marus, *The Nuremberg War Crimes Trials, passim*.

17 E.R. Fidell, *Military Justice* (Oxford: Oxford University Press, 2016); A. Duxbury and M. Groves, eds, *Military Justice in the Modern Age* (Cambridge: Cambridge University Press, 2016); G.S. Corn et al., eds, *The War on Terror and The Laws of War* (Oxford: Oxford University Press, 2015).

18 D. Orentlicher, *Some Kind of Justice: The ICTY's Impact in Bosnia and Serbia* (Oxford: Oxford University Press, 2018); S. Brammertz and M. Jarvis, *Prosecuting Conflict-Related Sexual Violence at the ICTY* (Oxford: Oxford University Press, 2016); V. Peskin, *International Justice in Rwanda and the Balkans* (Cambridge: Cambridge University Press, 2008); G. Gahima, *Transitional Justice in Rwanda* (London: Routledge, 2012).

19 W. Schabas, *An Introduction to the International Criminal Court* (Cambridge: Cambridge University Press, 5th edn, 2017); G. Solis, *The Law of Armed Conflict* (Cambridge: Cambridge University Press, 2nd edn, 2016).

20 D. Archibrugi and A. Pease, *Crime and Global Justice* (London: Polity, 2018); D. Bosco, *Rough Justice* (Oxford: Oxford University Press, 2014).

21 Archibrugi and Pease, *Crime and Global Justice, passim*; Schabas, *An Introduction to the International Criminal Court, passim*. Many thanks to George Owers for the "loser's justice" locution.

22 B. Orend, *The Morality of War* (Peterborough, ON: Broadview, 2nd edn, 2015), 190–1.

23 Perhaps no one has made a better case for both accountability and proportionality in settlements than Larry May, e.g., in his *After War Ends* (Cambridge: Cambridge University Press, 2012).

24 Orend, *The Morality of War*, 185–90.

25 R. Nozick, *Anarchy, State and Utopia* (Cambridge, MA: Harvard University Press, 1974).

26 P. Stearns, ed., *Demilitarization in the Contemporary World* (Chicago, IL: University of Illinois Press, 2013); D. Cummings, *Demilitarization* (London: Biblioscholar, 2012).

27 B. Orend, "Justice after War," *Ethics and International Affairs* (2002), 43–56. This may be a condition where it's debatable what's more

punitive: leaving the locals to their own devices and offering no help, or forcing regime change. I believe the former.

28 M. Neiberg, *The Treaty of Versailles* (Oxford: Oxford University Press, 2017); M. MacMillan, *Paris 1919: Six Months That Changed the World* (New York: Random House, 2003); D. Andelman, *A Shattered Peace* (London: Wiley, 2nd edn, 2014).

29 W. Danspeckgruber and C. Tripp, eds, *The Iraqi Aggression Against Kuwait* (Boulder, CO: Westview, 1996); J.T. Johnson and G. Weigel, eds, *Just War and Gulf War* (Washington, DC: University Press of America, 1991).

30 A. Arnove and A. Abunimah, eds, *Iraq Under Siege* (London: South End Press, 2000); G. Simons, *The Scourging of Iraq* (New York: Macmillan, 2nd edn, 1996).

31 E. Karsh and I. Rauter, *Saddam Hussein: A Political Biography* (New York: Grove, 2003).

32 For fuller narratives on the cases, see: Orend, *The Morality of War*, 226–32. On Germany: E. Davidson, *The Death and Life of Germany: An Account of the American Occupation* (Columbia, MO: University of Missouri Press, 1999); B. Steil, *The Marshall Plan* (New York: Simon & Schuster, 2018). On Japan: H. Schonberger, *Aftermath of War: Americans and the Remaking of Japan* (Kent, OH: Kent State University Press, 1989); M. Schaller, *The American Occupation of Japan* (Oxford: Oxford University Press, 1987).

33 J. Dobbins et al., *America's Role in Nation-Building: From Germany to Iraq* (Washington, DC: RAND, 2003).

34 M. Tondini, *Statebuilding and Justice Reform: Post-Conflict Reconstruction in Afghanistan* (New York: Routledge, 2010); M. Lamani and B. Momani, *From Desolation to Reconstruction: Iraq's Troubled Journey* (Waterloo, ON: CIGI, 2010).

35 Dobbins, *America's Role in Nation-Building*, passim; J. Dobbins et al., *The United Nations' Role in Nation-Building: From the Congo to Iraq* (Washington, DC: RAND, 2005). J. Dobbins et al., *Europe's Role in Nation-Building: From the Balkans to Congo* (Santa Monica, CA: RAND, 2008); J. Dobbins et al., *The Beginner's Guide to Nation-Building* (Santa Monica, CA; RAND, 2009).

36 US Government Accountability Office (GOA), *Afghanistan Reconstruction: Despite Some Progress, Deteriorating Security and Other*

Obstacles Threaten Achievement of US Goals (Washington, DC: Books LLC, 2011); D. Zakheim, *A Vulcan's Tale: How the Bush Administration Mismanaged the Reconstruction of Afghanistan* (Washington, DC: Brookings Institution, 2011); and on Trump's authorization, see M.R. Gordon, "Trump Gives Mattis Authority to Send More Troops to Afghanistan," *The New York Times*, June 13, 2017. Available at: https://www.nytimes.com/2017/06/13/world/asia/mattis-afghanistan-military.html?_r=0.

37 US Special Inspector General, *Hard Lessons: The Iraq Reconstruction Experience* (Washington, DC: US Independent Agencies and Commissions, 2009); F. Gerges, *ISIS: A History* (Princeton, NJ: Princeton University Press, 2016); L. Martinez, "Thousands More US Military Service Members in Iraq and Syria Than Believed," *ABC News*, November 27, 2017. Available at: http://abcnews.go.com/International/thousands-us-military-service-members-iraq-syria-believed/story?id=51411555.

38 Lamani and Momani, *From Desolation to Reconstruction, passim.*

39 W. Malley, *Transition in Afghanistan: Hope, Despair, and the Limits of Statebuilding* (New York: Routledge, 2018).

40 J. Bridoux, *American Foreign Policy and Post-War Reconstruction: Comparing Japan and Iraq* (New York: Routledge, 2012). See also the excellent documentary *No End in Sight*.

41 S. Eizenstat, J.E. Porter, and J.M. Weinstein, "Rebuilding Weak States," *Foreign Affairs* (2005), 134–46.

42 C.T. Call, *Constructing Justice and Security After War* (Washington, DC: US Institute of Peace Press, 2007).

43 Eizenstat et al., "Rebuilding Weak States," 134–46; C.T. Call, "Beyond the Failed State: Towards Conceptual Alternatives," *European Journal of International Relations* (2010), 303–26.

44 Eizenstat, "Rebuilding Weak States," 134–46; R. Rotberg, *When States Fail* (Princeton, NJ: Princeton University Press, 2003); L. Zambernardi, "Counter-Insurgency's Impossible Trilemma," *The Washington Quarterly* (2010), 21–34.

45 B. Orend, *Human Rights: Concept and Context* (Peterborough, ON: Broadview, 2002); Call, *Justice and Security After War, passim.*

46 L. Peperkamp, *Jus Post Bellum and the Nature of Peace* (Oisterwijk: Wolf Legal, 2017). See also the important individual, and collective,

work of Carsten Stahn, Jennifer Easterday, and Jens Iverson, such as their two edited collections with Oxford University Press: *Jus Post Bellum: Mapping the Normative Foundations* (2014) and *The Justice of Peace and Jus Post Bellum* (forthcoming).

Chapter 7

1 D. Sorenson, *Syria in Ruins* (New York: Praeger, 2016); J. Keegan, *A History of Warfare* (New York: Vintage, 1994).

2 K. Dockery, *Future Weapons* (New York: Berkley, 2007); Human Rights Watch, *Report on US Blinding Laser Weapons* (New York: Human Rights Watch, 1995).

3 J. Galliott and M. Lotz, eds, *Supersoldiers: The Ethical, Legal and Social Implications* (New York: Routledge, 2017); J. Moreno, *Mind Wars: Brain Science and the Military in the 21st Century* (New York: Bellevue, 2012).

4 It's estimated that, since 2000, the amount of US spending on drones has risen from US$284 million to more than US$3.3 billion. See: J. Gertler, "US Unmanned Aerial Systems," study for the US Congressional Research Office (January 2012): www.state.gov/documents/organization/180677.pdf. See also B. Roggio and A. Mayer, "Charting the Data for US Air Strikes in Pakistan, 2004–2012," *Long War Journal* (October, 2012) and M. Martin and C. Sasser, *Predator: The Remote-Control Air War over Iraq and Afghanistan* (New York: Zenith, 2010).

5 C.J. Fuller, *See It, Shoot It: The History of the CIA's Lethal Drone Program* (New Haven, CT: Yale University Press, 2017).

6 B.J. Strawser, "Moral Predators," *Journal of Military Ethics* (2010), 342–68; J. Scahill, *The Assassination Complex* (New York: Simon and Schuster, 2016).

7 H. Gusterson, *Drone: Remote Control Warfare* (Cambridge, MA: MIT Press, 2016); see also https://www.csis.org/analysis/new-white-house-drone-report.

8 M. Ignatieff, *Virtual War* (London: Metropolitan, 2000); M. Benjamin, *Drone Warfare* (New York: Verso, 2013); P.J. Springer, *Military Robots and Drones* (New York: ABC-CLIO, 2013).

9 C. Enemark, *Armed Drones and the Ethics of War* (London: Routledge,

2013); P. Singer, *Wired for War* (New York: Penguin, 2009); Strawser, "Moral Predators," 342–68.

10 See the exchange between S. Shane, "The Moral Case for Drones" (www.nytimes.com/2012/07/15/sunday-review/the-moral-ca se-for-drones.html) and D. Brunstetter, "Can We Wage a Just Drone War?" (www.theatlantic.com/technology/archive/2012/07/ can-we-wage-a-just-drone-war/260055/). See also such recent documentaries as: *Drone* (2014); *National Bird* (2016); and *Unmanned: America's Drone Wars* (2013).

11 For the test swarm video, see: www.youtube.com/watch?v=DjUd VxJH6yI. For the Syrian swarm, see: www.cnbc.com/2018/ 01/11/swarm-of-armed-diy-drones-attacks-russian-military-base- in-syria.html. Sources in general: A. Barrie, *Future Weapons: Access Granted* (London: Arbuthnot & Trevelyan, 2016); Gusterson, *Drone*.

12 See B. Perrigo's extended April 2018 *Time* article on all related weapons systems, with neat in-links, at: time.com/5230567/killer-robots/. Consider as well the important work of distinguished academics Noel Sharkey and Stuart Russell, such as Russell's viral video *Slaughterbots* or Sharkey's BBC series *Robot Wars*.

13 F. Allhoff, A. Henschke, and B.J. Strawser, eds, *Binary Bullets* (Oxford: Oxford University Press, 2016); S. Brenner, *Cyber Threats* (Oxford: Oxford University Press, 2009).

14 J. Carr, *Inside Cyber-Warfare* (London: O'Reilly, 2nd edn, 2011); R. Clarke, *Cyber War* (New York: HarperCollins, 2010).

15 W.J. Lynn, "Defending a New Domain: The Pentagon's Cyberstrategy," *Foreign Affairs* (Sept./Oct. 2010), 97–108; B.R. Allenby, *The Applied Ethics of Emerging Military and Security Technology* (London: Routledge, 2016).

16 L. Floridi, *Information* (Oxford: Oxford University Press, 2010). On "WannaCry": http://www.bbc.com/news/world-us-canada- 42407488.

17 M. Gross, "Enter the Cyber-Dragon," *Vanity Fair* (September 2011), 220–34. On Trump's tariffs/sanctions: https://www.theguardian. com/world/2018/mar/22/china-us-sanctions-trade-war.

18 G. Merritt, *Slippery Slope: Brexit and Europe's Troubled Future* (Oxford: Oxford University Press, 2017); M. Nunes, *The Plot to Hack America* (New York: Skyhorse, 2016).

19 G. Lucas, *Ethics and Cyberwarfare* (Oxford: Oxford University Press, 2016); US Congressional House, *Computer Security: Cyber-Attacks and War Without Borders* (Washington, DC: Books LLC, 2011).

20 Nunes, *Plot to Hack America*, *passim*; WikiLeaks, *The WikiLeaks Files: The World According to US Empire* (London: Verso, 2015). B. Hyacinthe, *Cyber-Warriors at War* (New York: XLibris, 2011).

21 The Economist, "Special Report on Cyberwar: War in the Fifth Domain," *The Economist* (July 1, 2010), 18–26.

22 A. Karatzogianni, ed., *Cyber-Conflict and Global Politics* (London: Routledge, 2008); R. Clarke, *Cyber-War: The Next Threat to National Security and What to Do About It* (New York: Ecco, 2012).

23 See the full, free report, at: https://www.us-cert.gov/ncas/alerts/TA18-074A.

24 M. Gross, "The Stuxnet Cyber-Weapon," *Vanity Fair* (April 2011), 152–98; M. Gross, "The Fog of Cyber-War," *Vanity Fair* (April 2011), 155–98; Lucas, *Ethics and Cyberwarfare*, *passim*. There's also the VICE documentary TV series, *Cyberwar*, detailing dozens of further cases.

25 H. Dinniss, *Cyber-Warfare and the Laws of War* (Cambridge: Cambridge University Press, 2012).

26 M. Schmitt, ed., *The Tallinn Manual on The International Laws Applicable to Cyberwarfare* (Cambridge: Cambridge University Press, 2013). There is a "2.0" updated version as well.

27 See Allhoff et al., *Binary*, *passim*; R. Dipert, "The Ethics of Cyberwarfare," *Journal of Military Ethics* (2010), 384–410; M. Cook, "'Cyberation' and Just War Doctrine," *Journal of Military Ethics* (2010), 417–22; B. Orend, "Fog in the Fifth Dimension: The Ethics of Cyberwar," in L. Floridi and M. Taddeo, eds, *The Ethics of Informational Warfare* (New York: Springer, 2014), 3–23.

28 Lucas, *Ethics and Cyberwarfare*, *passim*; J. Richards, *Cyber-War: The Anatomy of the Global Security Threat* (London: Palgrave Macmillan, 2014).

29 Lucas, *Ethics and Cyberwarfare*, *passim*; P. Singer and A. Friedman, *Cybersecurity and Cyberwar* (Oxford: Oxford University Press, 2014).

30 S.L. Myers, *The New Tsar: The Rise and Reign of Vladimir Putin* (New York: Vintage, 2015); M. Goldman, *Petrostate: Putin, Power and the New Russia* (Oxford: Oxford University Press, 2010); P. Longworth,

Russia: The Once and Future Empire (New York: St Martin's Press, 2006).

31 A. Ostrovsky, *The Invention of Russia* (New York: Viking, 2016); M. Gorbachev, *The New Russia* (Cambridge: Polity, 2016); D. Schoen, *Putin's Master Plan* (London: Encounter, 2016).

32 Schoen, *Putin's Master Plan, passim*; Allenby, *Applied Ethics, passim*.

33 Schoen, *Putin's Master Plan, passim*; Merritt, *Slippery Slope, passim*. See also http://www.cbc.ca/news/world/britain-fake-news-unit-1.4500172.

34 D. Dinan, N. Nugent, and W.E Paterson, eds, *The European Union in Crisis* (London: Palgrave, 2017); I. Berend, *The Contemporary Crisis of the European Union* (London: Routledge, 2016).

35 K.H. Jamieson, *Cyberwar: How Russia Helped Elect Trump* (Oxford: Oxford University Press, 2018); Allenby, *Applied Ethics, passim*.

36 P. Rosenzweig, *Cyber-Warfare* (New York: Praeger, 2013); D. Betz and T. Stevens, *Cyberspace and the State* (New York: Routledge, 2012).

37 I. Kant, *On Perpetual Peace*, ed. B. Orend and trans. I. Johnston (Peterborough, ON: Broadview Press, 2015). This volume includes the famous essay (51–102) as well the perpetual peace plans of St. Pierre, Leibniz, Rousseau, and such unexpected others as William Penn and Jeremy Bentham (103–17).

38 Kant, *On Perpetual Peace*, 11–44 and 60–73.

39 M. Doyle, "Kant, Liberal Legacies, and Foreign Affairs," *Philosophy and Public Affairs* (1983), 205–35, 323–53.

40 M. Doyle, *Liberal Peace* (New York: Routledge, 2011); M. Brown, S.M. Lynn-Jones, and S.E. Miller, eds, *Debating the Democratic Peace* (Cambridge, MA: MIT Press, 1996).

41 D. Welch, *Justice and the Genesis of War* (Cambridge: Cambridge University Press, 1988).

42 J.J. Rousseau, quote at p. 113 of the Orend edition of Kant's *On Perpetual Peace*. Rousseau's original essay was "A Lasting Peace through the Federation of Europe," of 1756.